Department for children, schools and families

Building Bulleting 100: **Design for fire safety in schools**

Foreword

The publication of Building Bulletin 100 (BB 100) is a landmark in improving fire safety in schools. It brings together guidance on how to make schools even safer places for children to be in, with guidance on how to protect the continuity of their education.

BB 100 is a design guide which shows clearly how the requirements for life safety, contained in the Building Regulations, can be met in the design of a new school or an extension. Innovative design may need to employ fire safety engineering, a risk-based approach, and this guide explains what design teams should do. It also covers the principles of fire safety management and describes the fire protection measures that the designer should consider.

What makes this different from previous fire safety guides is that it stresses the importance of protecting the fabric of schools. While the number of school fires has decreased over recent years, they remain a major risk for schools. Each year around 1 in 20 schools experiences a fire and nearly 60% of school fires are started deliberately. The short-term effects of loss of facilities and equipment can be calculated, but the longer-term effects of loss of coursework, disruption of classes and lowering of morale are much harder to quantify. However, it is clear that a major fire is likely to disrupt a child's education for many months and could mean postponing tests and exams. BB 100 shows how to protect school buildings from fire damage. It also includes extensive guidance on the use of sprinklers and their importance as a weapon against arson.

My interest in the potential for sprinkler systems to help prevent the devastating impact that a fire can have in a school is longstanding. A series of studies on their use were carried out, and on 1 March 2007 I announced that it is now our expectation that all new schools will have sprinklers fitted. Any exceptions to this will have to be justified by demonstrating that a school is low risk and that the use of sprinklers would not be good value for money. To help people make the right decisions, I launched two new tools that were developed for us by the Building Research Establishment. One covers risk assessments, enabling an existing or proposed school to be ranked high, medium or low risk. The other is a cost benefit analysis, specifically covering the use of sprinklers. They are both included in the CD-Rom attached to this guide.

It is my firm belief that using the guidance in BB 100, together with the tools, will ensure that the schools we build and refurbish now will be safer and better protected than ever before.

Jim Knight, MP

Minister of State for Schools and Learners

Department for Children, Schools and Families

Executive Summary

This guide provides fire safety design guidance for schools in England and Wales. The guidance applies to nursery schools, primary and secondary schools, including sixth form colleges, academies and city technology colleges, special schools and pupil referral units.

The guide is intended for all those with an interest in fire safety in schools, but in particular designers, fire engineers, building control officers (or equivalent) and fire safety officers. Head teachers, governors, teaching staff and facilities and maintenance staff will find it of interest to underpin their role as fire safety managers.

Formal requirements for life safety are covered by national legislation (Building Regulations) and supporting technical guidance with respect to fire; the guide is largely based on Approved Document B (Fire Safety) to the Building Regulations. A degree of property protection is an implicit consequence of the measures necessary to protect life. However, property protection measures that will satisfy insurers will generally be more onerous in some aspects. It is the intention of this guide to address both the life safety needs and the property protection needs at the same time. This dual approach will allow designers to tailor their strategy to the location, use, and risks identified.

The guide acknowledges the important role of sprinklers. Sprinkler systems installed in buildings can significantly reduce the degree of damage caused by fire and can reduce the risk to life. On 1 March 2007, DCSF announced the new policy on sprinklers and their value as a measure against the risk of fire and arson. All new schools should have fire sprinklers installed except in a few low risk schools. The tools available to carry out such a risk assessment are discussed in this guide.

Section 1 introduces the document, and summarises the regulatory background and general guidance on fire safety in schools. Section 2 gives background information and general design guidance. Also outlined is the risk assessment to be used during the design process. This provides a way of selecting measures to control and manage those risks as a product of the likelihood and impact of the risk identified. It provides the basis for understanding why fires develop and spread and how smoke moves through the building. It then outlines how to minimise the growth and spread of the fire thus allowing pupils and staff to leave safely. Sections 3 – 8 provide detailed design guidance that, if followed, will usually enable the school design to satisfy the requirements B1 – B5 of the Building Regulations. Where further recommendations are made with regard to property protection, these are clearly identified as such. Appendices provide further detailed information to supplement the main text.

The principles of fire safety management are summarised so that the designer can seek to ensure that the built school can be managed safely and effectively.

A designer is not required to follow the guidance in this document, but may adopt an alternative approach, possibly based on fire safety engineering. This is a risk-based approach, with the aim of providing an acceptable level of safety that gives good value for money. The onus is on the designer to demonstrate that the design results in an appropriate safety level, as good or better than that achieved by following the detailed design guidance here.

Contents

Contents

Contents

Section 1
Introduction

1.1   Audience

This guide is intended for all those with an interest in fire safety in schools, but in particular designers, fire engineers, building control officers (or equivalent) and fire safety officers.

Head teachers, governors, teaching staff and facilities and maintenance staff will find it of interest to underpin their role as fire safety managers.

1.2   How to use this guide

Formal requirements for life safety are covered by national legislation (Building Regulations) and supporting technical guidance with respect to fire. A degree of property protection is an implicit consequence of the measures necessary to protect life. However, property protection measures that will satisfy insurers will generally be more onerous in some aspects.

It is the intention of this guidance to address both the life safety needs and the property protection needs at the same time. This dual approach will allow designers to tailor their strategy to the location, use, and risks identified.

For ease of reading:

- key points are highlighted by an orange background (as above); and

- property protection recommendations are clearly distinguished from those concerned with life safety, by using a blue background to highlight them.

- References shown as (ref) are included in full, in appendix I.

Section 2 of this document gives background information and general design guidance. Sections 3 – 8 provide detailed design guidance that, if followed, will usually enable the school design to satisfy the requirements B1 – B5 of the Building Regulations, 2000. Where further recommendations are made with regard to property protection, these are clearly identified as such. Appendices provide further detailed information to supplement sections 3 – 8.

Whilst guidance appropriate to each of the requirements B1 – B5 of the Building Regulations is set out separately in this document, many of the provisions are closely interlinked. For example, there is a close link between the provisions for means of escape (B1) and those for the control of fire growth (B2), fire containment (B3) and facilities for the Fire and Rescue Service (B5). Similarly there are links between B3 and the provisions for controlling external fire spread (B4) and between B3 and B5. Interaction between these different requirements should be recognised where variations in the standard of provision are being considered. Similarly, a higher standard under one of the requirements may be deemed appropriate in seeking to provide an enhanced level of property protection, or in seeking to aid and/or improve the fire safety management of the school. The guidance in the document as a whole should be considered as a package aimed at achieving an acceptable standard of fire safety.

A designer is not required to follow the guidance in this document, but may adopt an alternative approach, possibly based on fire safety engineering. This should be a risk-based approach, with the aim of providing an acceptable level of safety that gives good value for money. The onus is on the designer to demonstrate that the design results in an appropriate safety level, as good or better than that achieved by following the detailed design guidance here.

Whenever new schools or significant refurbishments leading to material alterations (section 1.3.2) are being planned, a risk assessment should be performed to determine whether or not sprinklers should be installed (see section 1.6). Further guidance on performing risk assessment, with some tools that may be helpful, is provided in this document (section 2.4 and appendix H) and the accompanying CD-Rom.

1.3   Scope

This guidance on fire safety design covers schools in England and Wales. The guidance applies to nursery schools, primary and secondary schools, including sixth form colleges, academies and city technology colleges, special schools and pupil referral units.

Sixth form colleges designated as Institutions of Further Education are covered by Approved Document B (AD B) but BB 100 provides useful supplementary guidance on the design of educational buildings for students up to the age of 19.

This guide covers compliance with requirements B1 to B5 of the Building Regulations 2000 (which is concerned with life safety) but also provides guidance on the design of school buildings to reduce arson and property loss through fire.

1.3.1   Limitations to the scope

In order to produce a more manageable document, some circumstances less common in schools have not been covered in detail. If any of these cases apply, the designer should consult AD B for guidance. As the 2006 revision of AD B was drafted on the understanding that schools would be covered by BB 100 (this document), schools are not listed in the Purpose Groups covered by AD B. In most cases the appropriate purpose group will be 5 (Assembly & Recreation), or 2a if residential accommodation is involved.

The following buildings are not covered by this document:

- Buildings which include residential accommodation.
- Buildings with a top floor higher than 18m.
- Buildings with a basement deeper than 10m.
- Car parks (including those as part of the school building).

1.3.2   Material alteration

A 'material alteration' is one which, at any stage of the work, results in a building being less satisfactory than it was before in relation to compliance with the requirements of Parts B1, B3, B4 or B5 of the Building Regulations. Regulation 4(1) requires that any building work carried out in relation to a material alteration complies with the applicable requirements of Schedule 1 to the Regulations, while Regulation 4(2) requires that once that building work has been completed, the building as a whole must comply with the relevant requirements of

Schedule 1 or, where it did not comply before, must be no more unsatisfactory than it was before the work was carried out.

1.3.3   Buildings of special architectural or historic interest

In some cases where alterations to existing buildings are planned, particularly in buildings of special architectural or historic interest, adherence to the detailed design guidance in this document might prove unduly restrictive. In such cases it would be appropriate to take into account a range of fire safety features, some of which are dealt with in this document and some of which are not addressed in any detail and to set these against an assessment of the hazard and risk peculiar to the particular case.

1.4   Legal requirements

1.4.1   Building Regulations

Since April 2001, all new building work in schools has been subject to approval under the Building Regulations.

The functional requirements B1 to B5 of Schedule 1 of the Building Regulations are as follows:

B1: To ensure satisfactory provision of means of giving an alarm of fire and a satisfactory standard of means of escape for persons in the event of fire in a building.

B2: To ensure fire spread over the internal linings of buildings is inhibited.

B3: To ensure the stability of buildings in the event of fire; to ensure that there is a sufficient degree of fire separation within buildings and between adjoining buildings; to provide automatic fire suppression where necessary; and to inhibit the unseen spread of fire and smoke in concealed spaces in buildings.

B4: To ensure external walls and roofs have adequate resistance to the spread of fire over the external envelope and that spread of fire from one building to another is restricted.

B5: To ensure satisfactory access for fire appliances to buildings and the provision of facilities in buildings to assist fire-fighters in the saving of life of people in and around buildings.

In this document, each of the Requirements is dealt with separately in one of sections 4 – 8. The Requirement is reproduced at the start of the relevant section, followed by an introduction to the subject, and the detailed design guidance.

Any building work which is subject to the requirements imposed by Schedule 1 of the Building Regulations should, in accordance with Regulation 7, be carried out with proper materials and in a workmanlike manner. Further guidance can be found in the Approved Document supporting Regulation 7 on materials and workmanship.

Regulation 16B requires that where building work is carried out which affects fire safety, and where the building affected will be covered by the *Regulatory Reform (Fire Safety) (RRO) Order 2005*, the person carrying out the work must provide sufficient information for persons to operate and maintain the building in reasonable safety. This information will assist the eventual owner/ occupier/employer to meet their statutory duties under the *Regulatory Reform (Fire Safety) Order*. Appendix G provides advice on the sort of information that should be provided.

1.4.2 Interaction with other legislation

Under the *Regulatory Reform (Fire Safety) Order 2005 (RRO)* implemented in October 2006, fire safety legislation has become simplified and a suite of guides has been prepared for different occupancies. These deal with the provision and management of fire safety by risk assessment in the whole range of existing buildings; the one for schools is covered in *Risk Assessment Guide for Educational Premises 2006 (RRO Guide)*.

The Department for Children, Schools and Families document *Health & Safety: Responsibilities and Powers* (DfES/0803/2001) clarifies responsibilities for schools under existing health and safety legislation. With schools that are maintained by the Local Authority (LA), responsibility for fire safety is usually shared between the authority, the governing body and the head teacher. The LA usually has responsibility for alarm systems and the structural fire integrity of buildings, while the governing body and the head teacher are responsible for the day-to-day running of the school and the management of all systems including those for fire safety.

All three parties must ensure that school premises comply with Regulation 17 of *The Education (School Premises) Regulations 1999*. This requires that every part of a school building, and of the land provided for a school, shall be such that the safe escape of the occupants in case of fire is reasonably assured.

Particular regard is given to:

- the likely rate at which flames will spread across exposed surfaces;
- resistance to fire of the structure and of the materials of which the structures are made and their properties; and
- the means of escape in the case of fire.

1.5 Inclusive design

Current accommodation needs of school buildings must address inclusion where a wide age range of pupils may attend full time and also during community use where the buildings may be used out-of-hours by many groups, regardless of disability, age or gender.

Buildings and their facilities should be accessible to all who may work or live in them or visit them, without discrimination and it is likely that just as the requirements of AD M *Access and facilities for disabled people* and BS 8300:2001 *Design of buildings and their approaches to meet the needs of disabled people* have been drafted in response to the needs of the physically and visually impaired, the needs of the most disadvantaged will continue to influence future design standards.

Whilst there are obligations under the *Disability Discrimination Act 1995* for service providers and employers to facilitate accessibility, there are particular issues attending pupils in wheelchairs and those with multiple difficulties, and Special or Complex Needs in terms of general circulation and safe movement through and around buildings.

Some children with Special Needs use larger wheelchairs with additional support frames and other mobility aids which will impact on door and corridor widths over and above those required by AD M and BS:8300:2001 for horizontal escape. Places of safety/refuges will need to be planned and provided (BS5588:Part 8) to accommodate attendants if necessary, until the person can be escorted out of the building.

Those with audio/visual impairment and others with language difficulties may not be able to hear or read warnings, therefore in addition to conventional emergency signage, emergency routes may need to be highlighted by colour and texture changes on walls and floors.

The fire safety aspects of the Building Regulations Approved Document Part B are extended to schools in this document and further extended in detail in BB 77, *Designing for Pupils with Special Educational Needs and Disabilities*, to ensure that fire safety measures incorporated into a building take into account the access needs of all.

1.5.1 Out-of-hours use

With the increasing use of school buildings for community, or part-time use throughout the day and evening, the designer will need to cater for extended use very carefully, particularly for a refurbishment scheme. The use of only part of the building for community use (where alternative evacuation routes may be locked for school security reasons) along with the need to train a Responsible Person for day and evening activity is also an issue that needs to be addressed.

The major impact of the extended use is that the occupants of the building during these periods are less familiar with the layout and facilities than the daytime pupils are. As a consequence the escape plans and provisions must accommodate this change in occupancy even down to whether designated assembly points are adequate. Thus extended use will probably require the provision of:

- increased lighting, especially emergency lighting and external lighting;

- changes in evacuation planning and operation;
- increased car parking with attendant concerns about opportunities for vandalism and arson; and
- changes in door hardware selection.

1.6 DCSF policy regarding sprinkler systems

Sprinkler systems installed in buildings can significantly reduce the degree of damage caused by fire and can reduce the risk to life.

On 1 March 2007, DCSF announced the new policy on sprinklers and their value as a measure against the risk of fire and arson. All new schools should have fire sprinklers installed except in a few low risk schools.

Although the provision of sprinklers is not a requirement of the Building Regulations, DCSF expects that the Education Authority, Funding Body or overall 'client' of the scheme, should request, as part of the Employer's Requirements, that a risk assessment be undertaken to assess the validity of providing sprinklers in the scheme.

To help clients, local authorities and design teams assess the level of risk and make the right decisions, the DCSF has developed two new practical aids. The first is an interactive fire risk assessment tool. DCSF expects that this risk analysis will always be carried out and new schools being planned that score medium or high risk using the risk analysis tool will have sprinklers fitted.

The second tool is a cost benefit analysis tool. This tool helps users decide whether sprinklers represent good value for money.

These tools are included in a CD-Rom with this publication. Version updates will be published by DCSF on their fire safety website www.teachernet.gov.uk/fire

See section 2.4, and appendix H later.

1.7 Property protection

Figures from the Department for Communities and Local Government (CLG) show that over 1300 school fires a year in England and Wales are attended by local authority Fire and Rescue Services. Around 60% of these are started deliberately. For more statistics on school fires, see section 2.2. It is necessary to greatly reduce the risk of fires occurring in schools and, when a fire does occur, reduce the risk of it spreading.

While the primary concern is for the safety of the users of school buildings, a fire can have a serious impact on children's education due to disruption and loss of course work. The important roles that schools play in the community mean that losses incurred as a result of fire can have particularly severe social consequences. As such, it is important that property protection be considered during the design and throughout the working life of these buildings. This guide therefore gives advice on property protection as well as life safety issues.

For the purposes of this guide, the objectives of property fire protection include:

- minimising the effects of fire on the operation of the school (primarily, teaching);
- limiting the effects of interruption to operation of the school;
- seeking to have the school operational within 24 hours; and
- protecting the buildings.

While all of these objectives can be achieved only if the damage to the school from fire or smoke is minimised, it is also necessary for contingency plans to be in place to restore operations.

The insurance industry, Zurich Municipal, the Arson Prevention Bureau and the Arson Control Forum have produced various guides which address arson and vandalism and are directed at property protection (refs).

The principles recommended by the insurance industry for property protection are:

- construct the building from materials that will not contribute to a fire (except joinery);
- prevent premature collapse or excessive deflection;
- construct the building to minimise fire and smoke spread and confine the fire to its source;
- seek to minimise the risk of arson;
- design the building to avoid fire spread from another building or an external fire;
- provide appropriate automatic fire detection and alarm systems;
- ensure safety systems are maintained;
- ensure an adequate standard of fire safety management;
- seek to minimise the potential damage from fire-fighting water to the building (and the environment);
- ensure that all appropriate products or systems are third-party approved;
- ensure that all installers are third-party approved; and
- ensure that all heat-producing equipment, or equipment which might otherwise start a fire, is properly designed, installed and maintained.

In addition most insurers would recommend providing sprinklers for property protection.

Further information can be obtained from the FPA website: www.thefpa.co.uk

Many insurers use the Fire Protection Association's (FPA) *Design Guide for the fire protection of buildings* (ref) as a basis for providing guidance to the building designer on what they require. Some of the ways that the designer can reduce the effects of a fire using passive fire protection are by:

- limiting the use of easily ignited materials in the construction;
- using fire-resisting elements of construction (loadbearing and non-loadbearing);
- using smoke restricting elements of construction;
- providing features that limit the likely spread of flames and smoke production;
- providing features that prevent or limit fire and smoke exploiting cavities;
- preventing fire from exploiting services or ventilation ductwork; and
- limiting the potential for spread of fire to an adjacent building.

1.8 Alternative approaches – 'fire safety engineering'

Fire safety engineering can provide an alternative approach to fire safety. It may be the only practical way to achieve a satisfactory standard of fire safety as school building usage becomes ever more flexible in response to the needs of the local community. Fire safety engineering may also be suitable for solving a problem with an aspect of the building design which otherwise follows the provisions in this document.

Fire safety engineering is a risk-based approach and should provide a flexible way of solving the design problems and management issues. The fire safety strategy for the building may also include the level of management needed.

British Standard BS 7974 *Fire safety engineering in buildings* and supporting Published Documents provide a framework and guidance on the design and assessment of fire safety measures in buildings. Following the discipline of BS 7974 should enable designers and Building Control Bodies to be aware of the relevant issues, the

need to consider the complete fire-safety system and to follow a disciplined analytical framework.

Design teams must allocate time to consult with enforcing bodies and property insurers at an early stage to balance the requirements for life safety and recommendations for property protection.

For further information, refer to section 2.3.

1.9 Independent schemes of certification and accreditation

The performance of a system, product, component or structure is dependent upon satisfactory site installation, testing and maintenance. Independent schemes of certification and accreditation of installers and maintenance firms will provide confidence in the appropriate standard of workmanship being provided.

Confidence that the required level of performance can be achieved will be demonstrated by the use of a system, material, product or structure which is provided under the arrangements of a product conformity certification scheme and an accreditation of installers scheme.

Third party accredited product conformity certification schemes provide a means of identifying materials and designs of systems, products or structures which have demonstrated that they have the requisite performance in fire. In addition they provide confidence that the systems, materials, products or structures actually supplied are provided to the same specification or design as that tested/assessed.

Third party accreditation of installers of systems, materials, products or structures provides a means of ensuring that installations have been conducted by knowledgeable contractors to appropriate standards, thereby increasing the reliability of the anticipated performance in fire.

Building Control Bodies may accept the certification of products, components, materials or structures under such schemes as evidence of compliance with the relevant standard.

Similarly, Building Control Bodies may accept the certification of the installation or maintenance of products, components, materials or structures under such schemes as evidence of compliance with the relevant standard. Nonetheless, a Building Control Body will wish to establish, in advance of the work, that any such scheme is adequate for the purposes of the Building Regulations.

Insurers may require independent schemes of certification and accreditation of installers of a system, product, component or structure for the purposes of property protection.

Many certification bodies which approve such schemes are accredited by UKAS.

1.10 Fire safety management

Building Regulations do not impose any requirements on the fire safety management of a building. However, in developing an appropriate fire safety design for a building it may be necessary to consider the way in which it will be managed. A design which relies on an unrealistic or unsustainable management regime cannot be considered to have met the requirements of the Regulations.

This guidance has been written on the assumption that the building concerned will be properly managed. The use of the school will need to be flexible to respond to the changing needs of the occupants. A clear fire strategy report must be available on site; this will incorporate the management principles of the building with respect to fire.

Once the building is in use the management regime should be maintained and any variation in that regime should be the subject of a suitable risk assessment. Failure to take proper management responsibility may result in the prosecution of an employer, building owner or occupier under legislation such as the *Regulatory Reform (Fire Safety) Order 2005*.

1.10.1 The DCSF risk management strategy for schools and local authorities

In May 2007, DCSF launched a risk management strategy for existing schools and local authorities. It aims to improve risk management, school safety and security and help schools save money on their insurance premiums. The strategy includes a diagnostic risk ranking tool that enables local authorities to assess levels of risk in existing schools. Using this data, authorities can work towards weighting or discounting school insurance premiums as a financial incentive for schools to manage their risks. A risk management toolkit complements the risk ranking database and weighted premium approach. See www.dfes.gov.uk/riskmanagement or www.teachernet.gov.uk/riskmanagement

Section 2
Background information

Section 2 of this document gives background information and general design guidance.

Section 2.1 provides the basis for understanding why fires develop and spread and how smoke moves through the building. It then outlines how to minimise the growth and spread of the fire thus allowing pupils and staff to leave safely. Section 2.2 provides statistical data related to school fires, which helps to give an understanding of the scale of the problems. This information will also be directly relevant to fire risk assessments.

A designer is not required to follow the detailed design guidance in sections 3 – 8, but may adopt an alternative approach, possibly based on fire safety engineering. This is a risk-based approach, with the aim of providing an acceptable level of safety that gives good value for money. The onus is on the designer to demonstrate that the design results in an appropriate safety level, as good or better than that achieved by following the detailed design guidance. An outline of the fire safety engineering process is given in section 2.3, and risk assessment and cost-benefit analysis are both briefly covered in section 2.4.

A design based on fire safety engineering may involve various protection methods and fire safety systems. These are briefly described in section 2.5, with cross-reference to other sections of this document and pointers to further information.

Arson is a particular problem for schools. Many arson attacks will be opportunistic and therefore whilst the removal of likely fuel, rubbish, etc will reduce the possibility of an ignition occurring, even the best design will not counter the efforts of the really determined arsonist. There are several strategies for making their efforts less effective. These are discussed in section 2.6.

The guidance in sections 3 – 8 has been written on the assumption that the building concerned will be properly managed. The principles of fire safety management are summarised in section 2.7 so that the designer can seek to ensure that the built school can be managed safely and effectively.

2.1 The principles of fire behaviour

2.1.1 How do fires start?

The Fire Triangle (figure 1) represents the three elements needed for a fire to develop, namely a fuel which can be ignited by a heat source in the presence of oxygen from the air. Control of the fire is by eliminating one of the three elements. Identifying fire hazards includes recognising the fuel load and whether there are any heat sources present, eg, in kitchens and laboratories.

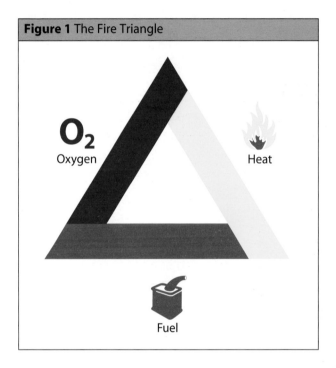

Figure 1 The Fire Triangle

O_2
Oxygen

Heat

Fuel

A fire can start when a source of (sufficient) heat is brought into contact with a combustible material (or flammable liquid) in the presence of air (oxygen).

Sources of heat include smokers' materials (in particular, lit matches, but also smouldering cigarettes), other naked flames (such as candles or gas cooker flames), overheating or sparking electrical equipment, chemical reactions, overheating motors, lightning, focussed solar rays, and many others.

In schools, combustible materials abound; such as paper, cloth and furnishings. Kitchens, workshops and laboratories may contain flammable liquids (such as cooking oil) or chemicals. Air is, of course, all around us.

2.1.2   Where do fires start?

Fire can start anywhere. Accidental fires can start when faulty over-heating electrical equipment, for example, is in contact with combustible material. They can therefore start anywhere, but they usually start inside a room. They should seldom start on escape routes, since fire safety management should ensure that these are kept clear.

Deliberate (arson) fires can be started anywhere, and often in places out of view, both inside and outside the building.

2.1.3   How do fires develop and spread?

When the fire starts in an enclosed space, hot smoke-laden gases will rise to the ceiling and form a layer which will flow under the whole ceiling at first and if not controlled it will then deepen and eventually fill the whole space. This is known as 'smoke-logging'. As the fire grows in area, flames and radiated heat will spread to any combustible materials such as fittings, furniture, exposed papers, etc. The flames will grow in length, increasing in height until they reach the ceiling where they will be deflected horizontally. Heat will then be radiated downwards which accelerates the growth of the fire as other items become involved in the fire leading to full involvement, ie, flashover. If there is little fresh air getting into the space the rate of fire development will slow as there is less oxygen present to keep the fire going. The gases generated will be very toxic and high in carbon monoxide.

This smoke will be irritant, asphyxiant and toxic, causing coughing, streaming eyes and difficulty in breathing. It will also be dense, and will restrict vision rapidly, resulting in disorientation and difficulty in moving away from the fire-affected area. The radiation from the flames once they reach the ceiling will promote fire growth. Any material which is burning will be hot and will use up the oxygen present and will produce carbon monoxide which will also disorient anyone in the area.

So, a fire starting in a compartment in a building may not only put anyone present in the room of origin at risk from the effects of the combustion products but if uncontrolled it will spread to other parts of the building as well. This could jeopardise the safety of people remote from where the fire started and could also cause damage over a wide area.

Thus schools should be designed, maintained and managed to reduce fire spread.

2.1.4   Smoke movement and its impact on escape

Early on in the fire, the most critical effects on the occupants will be from the smoke and other products of combustion. Smoke is often the first thing to be noticed by the occupants and is generally the cause of the first alarm. In the absence of any strong air currents smoke tends to gather at ceiling level, filling the space from the top down, ie, smoke logging. Once down to head height people will not be able to see very far because of the density of the smoke and the unpleasant effect on their eyes. Breathing will be difficult, as the carbon dioxide levels increase so the breathing rate goes up and more smoke is taken in which will be increasingly short of oxygen. The effects will be felt very quickly by younger pupils, who breathe quite rapidly anyway; leading to respiratory distress because the smoke is hot and eventually to unconsciousness or death as a result of carbon monoxide poisoning.

Recognition of these circumstances is essential when dealing with different age groups who may be familiar with the building but may not be very mobile for a variety of reasons. Legislative controls are usually cast with adults in mind. Although compliance can be achieved for adult occupants, designers and users of school buildings do need to consider the younger, smaller and shorter pupils at the planning stage to allow everyone, regardless of age, size and disability to leave safely.

It is important, therefore, that the escape routes must be protected against smoke penetration and the storey or final exit must be reached before they become untenable. Once visibility has dropped below 10m it will be difficult to move safely to the exits. The time taken will also need to take account of such things as the furniture present and the size of exits and the age and mobility of the occupants.

Thus all the right measures must be in place to allow safe egress from any part of the building.

2.2 Statistical data for school fires

DCSF intend to provide links to updated statistical information on school fires (such as that presented below), via their fire safety website.

The estimated number of accidental and non-accidental fires attended by Local Authority Fire and Rescue Services between 1993 and 2005 is shown in figure 2 (source: Department for Communities and Local Government (CLG) Fire Statistics).

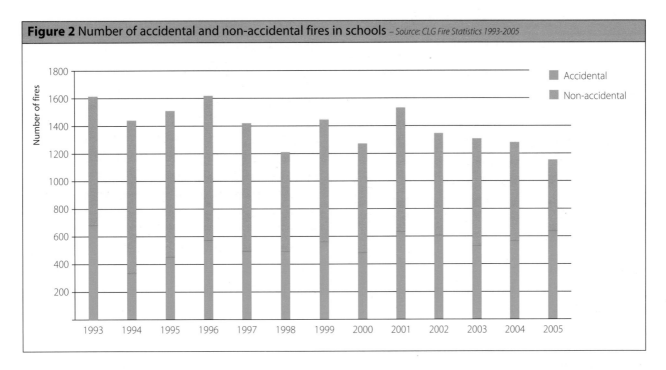

Figure 2 Number of accidental and non-accidental fires in schools – *Source: CLG Fire Statistics 1993-2005*

2.2.1 Deaths

The number of deaths in school fires is very small.

During the period 1994-2002, only one person died, in approximately 14,700 fires. The casualty was a 56-year old man, who died from burns. He was found in the place of fire origin (the roof space) where he had been working with a blow lamp (or similar heat source). The fire occurred at about 14:45 on Saturday 22 February 1997 (Gamble 1994-2002).

Despite the statistics, it is important to avoid complacency; there is always the possibility of a death in a school fire or even multiple deaths.

2.2.2 Injuries

The number of injuries in school fires is also quite small. During the period 1994-2002, 461 people were injured, an average of 51 per year, and 0.03 injuries per fire. Not all injuries are severe. In 2002, there were 46 injuries, as follows:

- 14 people suffering from smoke inhalation;
- 5 people suffering from burns;
- 4 people suffering from physical injuries (cuts, sprains, abrasions, etc);
- 2 people suffering from shock;
- 2 people suffering from other injuries; and
- 19 people referred to hospital for precautionary checks.

It is possible to put the injury rate per fire into context, by considering different building types. The worst fires are in dwellings which is not that surprising. Schools and Further Education (highlighted by red arrows in figure 3) are low risk (for life safety) compared to other building purpose groups. Some other purpose groups that one would not consider high risk, eg, supermarkets and restaurants (highlighted by green arrows in figure 3), are worse than schools.

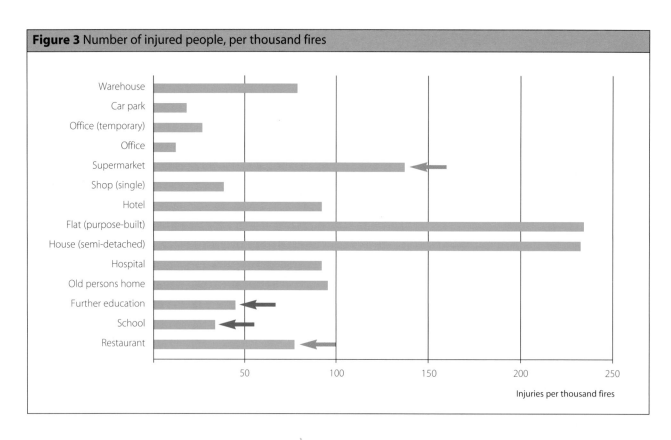

Figure 3 Number of injured people, per thousand fires

Injuries per thousand fires

2.2.3 Economic cost

A major issue with fires in schools is the scale of the property losses. The losses are a significant and constant drain on resources. According to government estimates (CLG), the average annual cost of school fires 2000-2004 was £58 million per year. Figures for the losses provided by the insurance industry are even higher.

Also note, this is just the direct losses (insurance claims) – the true cost is likely to be even higher. It is worth considering the kinds of uninsured losses that may occur:

- The hire of temporary accommodation during the rebuild – can be up to five years.
- Additional costs if pupils need to be transported to another school site.
- General disruption, including children's education.
- Loss of teaching aids.
- Loss of coursework – which will have to be redone by the pupils.
- Loss of personal items owned by pupils and staff.
- Loss of facilities for the community, eg, for Scouts, Guides, football and other sports,

evening classes, polling station, council meetings, centre for emergency accommodation.
- Additional costs from insurers requiring more security to prevent repeat incidents.
- Stress will be high, particularly for senior staff.
- Loss of reputation with implications for recruitment and retention of staff as well as pupil applications.

In context, schools top the list in terms of fire losses compared to other occupancies.

- In 2003 and 2004, educational fire losses were twice the next highest occupancy type (retail).
- In 2004, educational fire losses were 60 times those of office fires (the occupancy often referred to for future design comparisons).

In secondary schools where fires are a frequent occurrence, 43% report a fire occurring in one year with the average loss being £100,000 (CLG Research Bulletin No 10 – Survey of school fires 2006).

The cost in terms of lost investment in new and refurbished buildings together with 'interruption to business' and the wrecking of pupil's work and staff resources cannot be overstated.

Figure 4 Number of accidental and non-accidental fires occurring at different times of day

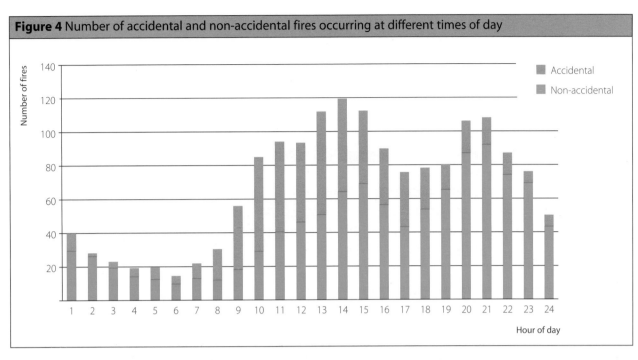

Figure 5 School fires per million population for different Fire and Rescue Service areas

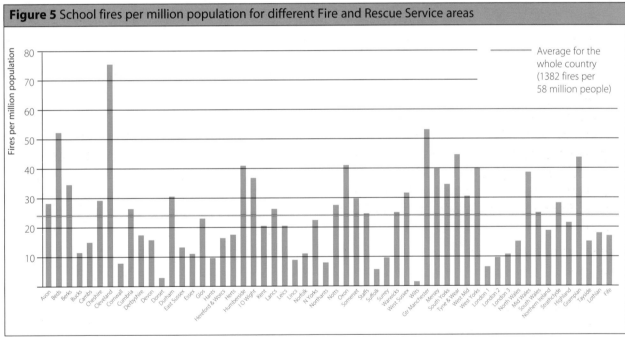

2.2.4 Pattern of accidental and deliberate fires

According to CLG statistics for the years 2000-2005, there have been just over 1,300 fires per year on average over this period. 56% of these fires were classified as 'non-accidental', with the majority of these likely to have been arson.

Received wisdom was that deliberate fires occur when the school is unoccupied so there is a low risk to life apart from to those lighting the fires and to fire-fighters. However, the perception is that this is now changing, with up to a third of

deliberate fires starting during the school day. The risk to pupils can be high with the younger pupils dependent on teachers and other staff for their safety in the event of fire. Typically, however, the larger fires still occur out of hours when the school is closed and the fire setter can work undisturbed and/or there is a delay in discovery.

Figure 4 shows the pattern of accidental and non-accidental fires during the day. This trend has remained fairly constant over the years studied, which suggests that the perception has only recently come into line with the reality.

2.2.5 Relative number of school fires in different regions of the country

Figure 5 shows the number of school fires per million head of population, for each of the Fire and Rescue Service areas in the UK. Note that there is considerable variation between different regions. CLG (Fire Statistics, United Kingdom, 2004, CLG February 2006) indicate that 40% of school fires occur in the metropolitan areas.

This factor is one of those that should be considered when performing a risk assessment for a particular school project.

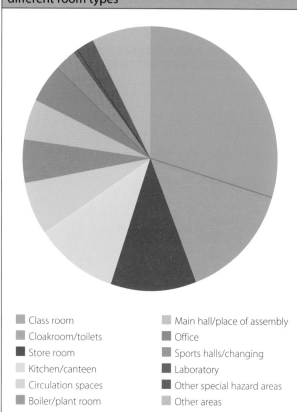

Figure 6 relative number of reported fires starting in different room types

- Class room
- Cloakroom/toilets
- Store room
- Kitchen/canteen
- Circulation spaces
- Boiler/plant room
- Main hall/place of assembly
- Office
- Sports halls/changing
- Laboratory
- Other special hazard areas
- Other areas

2.2.6 Fire locations within a school

The relative number of reported fires (CLG fire statistics 2002) starting in different room types is shown in figure 6. Note that some areas of special fire hazard (eg, heat bays) do not appear within this pie chart, partly because such rooms are relatively few in number, and partly because extra fire safety precautions (eg, extinguishers) will have been provided, enabling fires to be dealt with while they are still small, and thus not

requiring Fire and Rescue Service intervention. Note the relatively large number of cloakroom and toilet fires, most of which are arson.

This information can be used as part of a quantified risk assessment, to estimate the frequency of fires starting in different room types.

2.3 Fire safety engineering

The definition of fire safety engineering used by the Institution of Fire Engineers is:

'The application of scientific and engineering principles based on an understanding of the phenomena and effects of fire and of the behaviour of people to fire, to protect people, property and the environment from the destructive effects of fire.'

Other organisations, such as ISO, have published similar definitions (see, eg, ISO TR 13387).

The principal objective of fire engineering is, when fire occurs, to provide an acceptable level of safety. Often this will involve calculation or modelling of scenarios affecting all or part of the fire 'system'.

Fire safety engineering is a risk-based approach and should provide a flexible and cost-effective way of solving the design problems and management issues. It requires taking a more holistic approach to a problem than may be taken when using more prescriptive methods. Additional factors, covered implicitly in the prescriptive approach, need to be considered explicitly.

It is often the interactions between different aspects that cause difficulties in practical situations. Solutions to problems of a building's day-to-day use (ventilation, structural, security, etc) may conflict with requirements for fire safety. Fire safety engineering must be undertaken using a systematic approach to avoid potentially life-threatening omissions in the analysis. The flow diagram (figure 7) outlines a suggested approach to the design process, following BS7974 (2001): *Application of fire safety engineering principles to the design of buildings – Code of practice.*

Fire safety engineering will be most effective if it is included in the overall design process at the earliest possible opportunity. It should also be emphasised that the process is iterative, continuing throughout the overall design process until the design is complete.

The key stages are:

- the qualitative design review (QDR);
- the quantitative analysis;
- the assessment; and
- reporting.

Here, particularly at the qualitative design review stage, the interaction of all the building systems are considered as well as the detailed performance of the fire protection systems. A wide variety of measures could be considered and incorporated to a greater or lesser extent, as appropriate in the circumstances.

These include:

- the adequacy of means to prevent fire;
- early fire warning by an automatic detection and warning system;
- the standard of means of escape;
- the standard of active measures for fire extinguishment or control;
- provision of smoke control;
- control of the rate of growth of a fire;
- structural robustness and the adequacy of the structure to resist the effects of a fire;
- the degree of fire containment;
- fire separation between buildings or parts of buildings;
- facilities to assist the Fire and Rescue Service;
- availability of powers to require staff training in fire safety and fire routines;
- consideration of the availability of any continuing control under other legislation that could ensure continued maintenance of such systems; and
- management.

The QDR, assessment and reporting are discussed in the following sections. The quantitative analysis is not covered here, as the range of possible calculations is too wide (it is the role of the QDR to decide which calculations are appropriate for a given project). The assumptions made when quantitative methods are used need careful assessment.

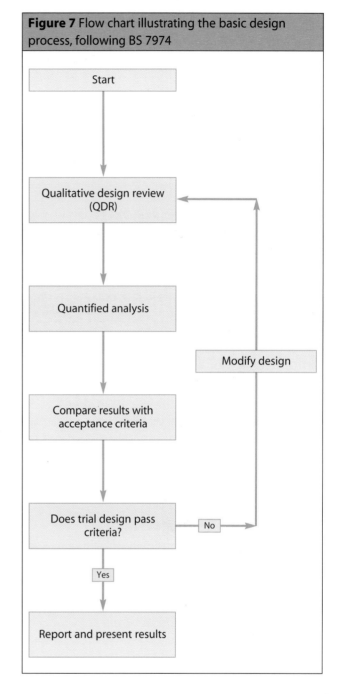

Figure 7 Flow chart illustrating the basic design process, following BS 7974

2.3.1 Qualitative design review

The fire safety design process begins with the qualitative design review (QDR). During this stage the scope and objectives of the fire safety design are defined, and performance criteria are established. One or more potential designs will be considered, for a detailed quantitative assessment. If the proposed design(s) are unsatisfactory according to the performance criteria chosen, the QDR must be repeated and new or modified designs will need to be considered.

Tasks carried out in the QDR include:

- review of the architectural design;
- determine building, environment and occupant characteristics;
- establish fire safety/protection objectives;
- decide acceptance criteria;
- identify fire hazards and possible consequences;
- propose trial fire safety/protection designs;
- propose an evacuation strategy;
- make reasonable assumptions to simplify the problem;
- specify scenarios for analysis;
- indicate appropriate methods of analysis; and
- report the results of the QDR stage.

For large projects, the QDR should be performed by a group of people, including members of the design team, one or more fire safety engineers, and others as appropriate (such as Building Control Bodies and representatives of the fire service and insurers). The results of the QDR should be included in the final design report seeking approval of the design.

2.3.2   Assessment of designs

In a fire safety engineering design following the procedures given in BS7974, the assessment of a design requires the results of a quantified analysis to be compared with the original design criteria determined during the qualitative design review (QDR). The assessment method and criteria should be agreed before the quantitative design stage.

Factors that should be taken into account include:

- the anticipated probability of a fire occurring;
- the anticipated fire severity;
- the ability of a structure to resist the spread of fire and smoke; and
- the consequential danger to people in and around the building.

BS7974 gives three methods of assessing a design:

- Comparative criteria where a design is shown to have a comparable level of safety to some other method (eg, Approved Document B, for buildings in England and Wales).
- Deterministic criteria where a specific set of safety conditions are shown to have been achieved (eg, smoke layer is never below a specified height). The QDR should establish a worst case fire scenario (or scenarios) for the particular school in question and recognise that any uncertainty in the calculations should be considered. It is better to err on the side of safety if there are any doubts.
- Probabilistic criteria where the frequency of an event occurring (ie, probability of occurrence during a given time period) is shown to be less than an agreed small value. The consequences of the event should also be considered. The risk of an event is defined by multiplying the consequence by the frequency.

Failure of a design to meet the assessment criteria will require the design to be modified, and the QDR and quantitative analysis cycle to be repeated until a successful design has been determined. A design should not be tailored too closely to the minimum performance requirements, as buildings change over time and the fire strategy may have to be adapted as a result.

Trade-offs of one system for another need to be carefully evaluated as they may compromise not only overall fire safety provisions, because of reductions in the levels of safety back up and engineering redundancy, but also other building functions. Reductions in the level of compartmentation, for example, may compromise performance in other respects (such as energy efficiency, acoustic insulation, multiple space utilisation, security, and privacy). The designer needs to take these wider considerations into account in the overall achievement of functional design and cost targets.

2.3.3　Reporting and presentation

The design of buildings using a fire safety engineering approach will be subject to review and approval and should be reported so that the procedures and assumptions can be readily understood by a third party. Thus, assumptions and engineering judgements should be clearly identified, and sufficient detail should be included so that the quantified analysis can be repeated or reviewed by a third party.

BS 7974 does not provide a specific format for reporting and presentation as a prescribed format could not anticipate all the requirements of a performance based design. It does, however state that the following should be included:

- objectives of the study;
- building description;
- results of the qualitative design review;
- quantified analysis;
- comparison with acceptance criteria;
- fire safety strategy;
- management requirements;
- conclusions;
- references; and
- qualifications and experience of the fire safety engineer(s).

A sensitivity or uncertainty analysis should be performed to estimate the confidence limits for the key output variables that provide the comparison against the acceptance criteria.

2.4　Fire risk assessment

Any fire risk assessment should reflect the day-to-day use of the school as well as its design.

Risk assessment, either implicitly or explicitly, is a key part of building fire safety engineering. The final design of the building will present a way of dealing with the risks in a particular school, which have to be addressed during the life of that building. The risk assessment should not only examine the chances of an incident occurring but also the potential consequences of that incident, ie, the likelihood and impact assessed together. It is important to match the right risk assessment method to the decision to be made.

Implicit risk assessment examples include the comparison of calculation results with threshold criteria, eg, 'smoke layer well above people's heads' or 'area of fire spread restricted to less than X m²'; often these are linked with 'worst case' scenarios. The idea is that 'worst' and lesser scenarios lead to minimal consequences (once remedial or protective measures have been taken), with other more severe scenarios being assumed to have minimal probability. However, which scenario will be 'worst case'? Conservative assumptions for one aspect of the fire 'system' might not be conservative at all for other aspects. A sensitivity analysis should be performed to estimate the consequences of uncertainties in the scenario, variable values, etc.

Explicit risk assessment uses the formula:

$$Risk = \sum frequency \times consequence$$

where the summation sign applies to all hazards or scenarios. For example, suppose that the average damage, caused when a fire occurs, is £100,000. If a fire is expected to occur once every 10 years on average in a particular school (ie, a frequency of 0.1 year^{-1}), then the risk would be £10,000/year.

Every safety decision should require a full risk analysis, until or unless it can be shown that a less comprehensive approach is adequate. The preferred approach to uncertainty is to quantify it, rather than rely on conservative assumptions.

Risks may be reduced by preventing hazards from occurring (ie, reducing the frequency), from mitigating the consequences should hazards occur, or some combination of both.

No building can be completely safe, yet there is an unresolved question of what absolute level of risk should be deemed acceptable. One way in which these problems may be addressed is to make the assumption that the risks associated with buildings designed following prescriptive guidelines are considered to be 'reasonable'. Additionally, most quantitative methods are used in comparative mode, ie, there is an assumption that systematic errors and biases 'cancel out' when two similar designs are compared.

2.4.1 Risk assessment and the provision of sprinkler systems

On 1 March 2007 DCSF announced the new policy on sprinklers and their value as a measure against the risk of fire and arson. All new schools should have fire sprinklers installed except in a few low risk schools.

To help clients, local authorities and design teams assess the level of risk and make the right decisions, the DCSF has developed two new practical aids. The first is an interactive fire risk assessment tool. DCSF expects that this risk analysis will always be carried out and new schools being planned that score medium or high risk using the risk analysis tool will have sprinklers fitted.

This tool is included in a CD-Rom with this publication. Version updates will be published by DCSF on their fire safety website www.teachernet.gov.uk/fire

See appendix H later.

The question of whether a fire suppression system such as sprinklers should be fitted will depend on several factors all of which should be identified in the risk assessment. These factors include:

- probability of different fire scenarios;
- consequences of the fire scenarios;
- location of the buildings;
- how accessible they are;
- vulnerability to intruders through the perimeter of the site;
- whether there is public access to the site;
- vulnerability of the construction to fire involvement;
- capabilities of the security system;
- whether facilities for waste disposal and storage are well away from the buildings to prevent an external hazard coming into contact with the fabric of the building;
- whether there is previous history of vandalism and arson (existing schools only);
- how long it takes the Fire and Rescue Service to reach the buildings and fight the fire; and
- availability of water supply.

2.4.2 Cost benefit analysis (CBA)

CBA is a logical culmination of fire safety engineering; it enables decisions on optimising the balance between safety and cost to be made. Benefits and costs need to be expressed in the same monetary terms, for example £/year. An holistic approach should ideally be taken, encompassing life safety, property or asset protection, environmental impact, business continuity, etc, and the whole life of the building should be considered. For a provider of a PFI school, unavailability of facilities will be of particular concern. Some costs are incurred at the start of the project (eg, construction/refurbishment), others are ongoing (eg, maintenance). Benefits, in terms of deaths and injuries prevented, damage prevented or reduced, etc, would be ongoing, although factors such as reduced reliability of protection systems may also need inclusion.

Results of cost-benefit analysis can be presented as either the ratio between benefit and cost (ratio > 1 is good), or the difference (benefit − cost > 0 is good). As with any aspect of fire safety engineering, the uncertainty/confidence limits in the calculations need to be quoted, in order to interpret the significance of the result. There are four possible outcomes:

a) Proposed measure(s) clearly beneficial (ratio > 1 or difference > 0); confidence level ⩾ 95%. Note that strictly speaking the null hypothesis is 'measure is not cost effective', which the analysis 'disproves', ie, there is less than a 5% probability of the analysis incorrectly giving a significant result due to chance, when really the measure is not cost-effective.

b) Proposed measures clearly not beneficial (ratio significantly < 1 or difference significantly < 0), confidence level ≪ 50%.

c) Borderline (ratio ≈1 or difference ≈ 0), confidence level ≈ 50%, but with small uncertainty (similar or smaller magnitude than the result). In this case, if any factors have been omitted from the analysis, they may swing the decision one way or the other once they are included. Alternatively different measure(s) may need to be investigated.

d) Borderline, confidence level ≈ 50% but large uncertainty. Here the benefit:cost ratio may be ≫1, or the difference ≫0, but if the uncertainties are even larger, the cost-benefit analysis will be of no help in reaching a decision unless the uncertainties are reduced. Alternatively other measures could be investigated.

The cost-benefit analysis should also take into account any major knock-on effects of the proposed fire protection measures. It should consider the wider consequences, particularly if these measures have cost implications for the building as a whole regarding the overall functional value and sustainability of the building, including its operating effectiveness and efficiency.

In order to optimise cost-effectiveness, alternative solutions to the building design each require a cost-benefit analysis to be performed in order to select the best (see figure 8).

2.4.3 Cost-benefit analysis and the provision of sprinkler systems

The second tool included in the CD-Rom with this publication is a cost-benefit analysis tool. This tool helps users decide whether sprinklers or other approaches represent good value for money. Version updates will be published by DCSF on their fire safety website www.teachernet.gov.uk/fire

See appendix H later.

2.5 Fire protection systems

A design based on fire safety engineering may involve various protection methods and fire safety systems. These are briefly described below, with cross-reference to other sections of this document and pointers to further information.

For any of the systems mentioned here it is imperative that each one should be designed and installed to work at its optimum criteria. Further it is essential that the various systems do not impede the workings of any other system.

2.5.1 Means of escape

The design of means of escape from a building must be based on an appreciation of the probable behaviour of fire, which may break out in any part of the building and then spread to other parts. Although recommendations based on such considerations can be devised, they can be used intelligently only if the nature of the risks which they are intended to meet is continually borne in mind. The design of a building should therefore be analysed, part by part, in order to determine the danger which might arise from a fire, either in the part where the fire may originate or in any other part to which it may spread. The value of analysing a plan with these facts in mind cannot be over-stressed.

As mentioned earlier (section 2.1) the primary danger associated with fire in its early stages is

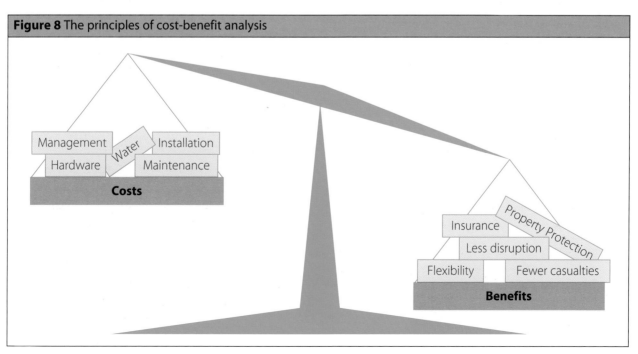

Figure 8 The principles of cost-benefit analysis

Management
Hardware
Water
Installation
Maintenance
Costs

Insurance
Property Protection
Less disruption
Flexibility
Fewer casualties
Benefits

not flame or heat but smoke and toxic gases produced by the fire. These may make an escape route impassable long before a temperature which is dangerous to life is reached. It is therefore important to ensure that the escape routes remain usable, ie, not blocked by smoke, for as long as required for people to evacuate the building.

The first and fundamental principle is the provision of alternative means of escape. Much of the guidance on means of escape in section 4 therefore revolves around this principle. It ensures that people should always be able to turn and walk away from a fire, except for very short distances at the start of their evacuation if they happen to be in close proximity to the fire.

One of the key ways the detailed design guidance ensures adequate means of escape is by setting upper limits on the travel distance to a storey or final exit. Whilst the primary effect of this is to limit the amount of time that people may potentially be affected by the fire before they reach the relative safety of a protected stair, or the ultimate safety of the final exit, there are also other implications.

Control of travel distance achieves the following:

- limited travel time; safety may be reached without serious exposure to smoke;
- limited size and complexity of enclosure;
- provision of sufficient alternative escape capacity within a reasonable distance. If there is a choice of exits, occupants should be able to escape in a direction away from the fire;
- increased likelihood that an exit is visible, and remains so during fire;
- reduced likelihood that a fire can occur unseen, or grow large before detection/alarm; and
- reduced likelihood of a fire between occupant and exit.

2.5.2 Automatic fire detection, alarms and communications

Although the staff and pupils may be expected to see a fire starting and/or smell the smoke this is not a reliable detection system. People are not present in all parts of the building nor are they there day and night, every day (24/7). Early automatic detection and alarm of the fire will allow occupants to escape quickly and safely, or tackle the fire while it is still at an early stage of development. It will enable professional help to be summoned without delay which should reduce the damage to the fabric and contents of the building(s).

Automatic fire detection may be linked to other systems in order to trigger their operation. Examples include hold-open devices for fire doors, smoke extraction or ventilation systems, suppression systems, and fire dampers in ventilation ducts.

Refer to section 3.2 for specific design benefits connected with the use of fire detection and alarm systems.

For guidance on system design, installation and servicing, BS 5839: Part 1: 2002 provides all the information a designer will need. Category P automatic systems are for property protection and Category L for life protection. Category M is the manual system providing break glass call points and sounders. Where break glass call points are provided, vandals can be deterred by fitting liftable covers with audio alarms.

A fully automated zoned fire detection system with a direct link to a central monitoring station is the best way to get rapid attendance by the Fire and Rescue Service, especially out of school hours. However in the light of the Fire Services Act 2004, this may need to be confirmed by a 999 call as the Act seeks to reduce attendance at false alarms.

False alarms can largely be eliminated at the design stage by careful choice of detector type, location of detectors, configuration of the system and linking it to the security system.

2.5.3 Signs and notices

Further guidance is given in RRO Guide Part 2, section 6 on safety signs and notices.

2.5.3.1 Signs

Signs must be used, where necessary, to help people identify escape routes, find fire-fighting equipment and emergency fire telephones. However, signs may not be required for very simple small premises where all the exits are in regular use and familiar to all (eg, in a small village school). Refer to section 4.5.5.6 for the signage requirements for escape routes.

These signs are required under the Health and Safety (Safety Signs and Signals) Regulations 1996 and must comply with the provisions of those Regulations. For a sign to comply with these Regulations it must be in pictogram form. The pictogram can be supplemented by text if this is considered necessary to make the sign more easily understood, but you must not have a safety sign that uses only text.

Appropriate signs should also take into account the age and ability of pupils or students.

Figure 9 Typical fire exit sign

2.5.3.2 Notices

Notices must be used, where necessary, to provide the following:

• instructions on how to use any fire safety equipment;

• the actions to be taken in the event of fire; and

• information for the Fire and Rescue Service.

All signs and notices should be positioned so that they can be easily seen and understood.

Figure 10 Staff action notice

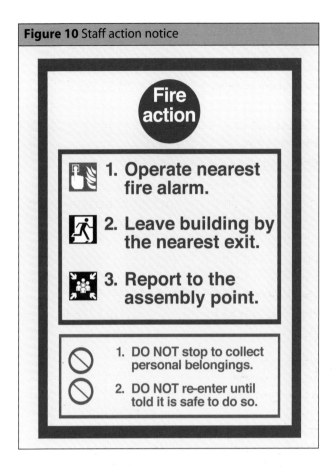

2.5.4 Emergency lighting

Emergency escape lighting (BS EN1838/BS 5266: Part 7: 1999 *Lighting applications: emergency lighting*) is required to fulfill the following functions:

1. Clearly indicate and illuminate escape routes and exit signs, including escape routes which are external to the building.

2. Ensure that changes of level and direction are indicated in accordance with Approved Document M of the Building Regulations.

3. Ensure that fire alarm call points and fire-fighting equipment can be easily located.

Detailed guidance is given in section 4.5.5.5.

In most cases it will be sufficient to have a system of 'non-maintained' emergency lighting, ie, where the luminaires are only illuminated if the normal lighting fails. If areas are to be used for activities which might require some form of licensing, eg, public entertainments licensing, then a system of 'maintained lighting' would be required in these areas. With a 'maintained' system the emergency luminaires are illuminated at all times.

Emergency lights can be powered by battery or a back-up generator; battery systems are more common and cost-effective in most school buildings but must be checked regularly. It is usual to find a small red LED indicating that the light is 'live' and will come on in the event of an emergency.

Further guidance is given in RRO Guide Part 2, section 5 on emergency lighting.

2.5.5 Smoke control

There are four main reasons for the control of the spread of smoke – to protect means of escape, to assist fire-fighting, to limit the risk to occupants in rooms not immediately in the vicinity of the fire and to minimise smoke damage to the contents and fabric of the building. In most cases this is achieved using containment measures such as doors and walls. However in some circumstances smoke ventilation systems or pressurisation systems may be necessary or desirable.

Smoke ventilation systems can extend the time for the presence of the smoke to cause life-threatening conditions as the space becomes smoke logged. It will depend on the volume of the space, the height of the ceiling and the size of the fire.

Some form of smoke ventilation of basements is required by the detailed design guidance to comply with the Building Regulations. This is to assist fire-fighters (section 8.5). In other parts of this detailed guidance, smoke control is optional for other purposes.

In addition to the threat to life safety described above, smoke will contaminate sensitive ICT equipment as well as books. Smoke control may therefore need consideration as a property protection system.

Where the design necessitates an engineered approach to smoke ventilation in large spaces, such as theatres, assembly halls and sports halls, advice on the design of such systems can be found in BRE Report BR 368, BS7346: Part 4:2003 or CIBSE Guide E. For specific building types, eg, a building containing an atrium, advice is contained in BS 5588 Part 7.

In any school building whether new or existing, the ventilation system is likely to involve some ductwork/louvres. Where these cross a compartment wall which has a designed fire resistance, that fire resistance must not be compromised.

A ventilation system that has been designed to clear any smoke from a fire will need a suitable detection system to actuate it.

2.5.6 First-aid fire-fighting equipment

The provision of first-aid fire-fighting equipment (eg, fire extinguishers) is not covered by the guidance to the Building Regulations (AD B). However, fire extinguishers may fulfil a life safety function if the safest way to escape from the effects of a fire is actually to put it out. Fire extinguishers also have an obvious role to play in property protection, if the fire can be tackled before it becomes too large.

Designers should give some thought to the provision of first-aid fire-fighting equipment, since this might involve creating storage niches in the walls, to give but one example. For more details on the numbers and types of fire extinguishers that might be required for different areas of the school building, refer to appendix D.

2.5.7 Sprinkler systems

Sprinkler systems installed in buildings can reduce the risk to life and significantly reduce the degree of damage caused by fire. Sprinkler protection can also sometimes be used as a compensatory feature where the provisions of this document are varied in some way. Where sprinklers are provided, it is normal practice to provide sprinkler protection throughout a building. However, where the sprinklers are being installed as a compensatory feature to address a specific risk or hazard, it may be acceptable to protect only part of a building. Further guidance can also be found in *Sprinklers for Safety: Use and Benefits of Incorporating Sprinklers in Buildings and Structures*, BAFSA 2006 (ISBN: 0 95526 280 1).

Sprinklers are known to be highly effective in controlling a fire while it is still small, and certainly will buy time before the arrival of the

Fire and Rescue Service, as well as creating less hazardous conditions for the fire fighters.

The sprinkler head acts rather like a thermometer, when it is immersed in hot smoke. The whole head heats up, and the liquid in the glass bulb expands until it shatters, whereupon water sprays out of the sprinkler delivering an umbrella-shaped spray of water droplets. The heads closest to the fire will normally heat up first, so when the water sprays out it should land on the fire and hopefully put it out. Note this is unlike the total involvement of sprinkler systems as misleadingly depicted in films and adverts. Even if the fire is not extinguished (eg, it is sheltered from the sprinkler spray), the wetting of the surrounding area should prevent the fire from spreading much further.

If a building-wide suppression system is installed it will have the added bonus that these will be heat detectors throughout the school which may lead to overall savings. Sprinkler systems incorporate flow meters connected to alarms (often with a direct link to the Fire and Rescue Service).

Recognised and recommended industry best practice is to employ sprinklers within a design concept of integrated fire protection, which applies a balance of different and complementary measures. Optimum sprinkler functionality also depends on full system maintenance, security of water supply and protection of feeder systems and pipes in the event of fire. Such considerations are an important part of any design risk assessment.

Detailed advice on the design of sprinkler systems for schools is given in the DCSF Standard Specification Layout and Design (SSLD) standard specification, *Sprinklers in schools*.

Simple industry guidance on the use of sprinklers can be found in the BAFSA guide, *Frequently Asked Sprinkler Questions – Information File 2005* (http://www.bafsa.org.uk/pdfs/publications/00000046.pdf).

Refer to section 3.3 for specific design benefits that may accrue from the installation of a sprinkler system.

On 1 March 2007 DCSF announced the new policy on sprinklers and their value as a measure against the risk of fire and arson. All new schools should have fire sprinklers installed except in a few low risk schools.

Although the provision of sprinklers is not a requirement of the Building Regulations, DCSF expects that the Education Authority, Funding Body or overall 'client' of the scheme, should request, as part of the Employer's Requirements, that a risk assessment be undertaken to assess the validity of providing sprinklers in the scheme.

To help clients, local authorities and design teams assess the level of risk and make the right decisions, the DCSF has developed two new practical aids. The first is an interactive fire risk assessment tool. DCSF expects that this risk analysis will always be carried out and new schools being planned that score medium or high risk using the risk analysis tool will have sprinklers fitted.

The second tool is a cost-benefit analysis tool. This tool helps users decide whether sprinklers represent good value for money.

These tools are included in a CD-Rom with this publication. Version updates will be published by DCSF on the Teachernet fire safety website www.teachernet.gov.uk/fire

See section 2.4 and appendix H.

2.5.8 Other automatic fire suppression systems

Fire suppression systems can cover the whole school, as in the case of sprinklers, or be provided to a specific area using systems such as gaseous or water mist systems which will target identified hazards.

There are many alternative or innovative fire suppression systems available. Where these are used, it is necessary to ensure that such systems have been designed and tested for use in buildings and are fit for their intended purpose. Most of these systems are not appropriate for use in schools.

If such systems are considered appropriate or necessary then reference should be made to the relevant British Standard.

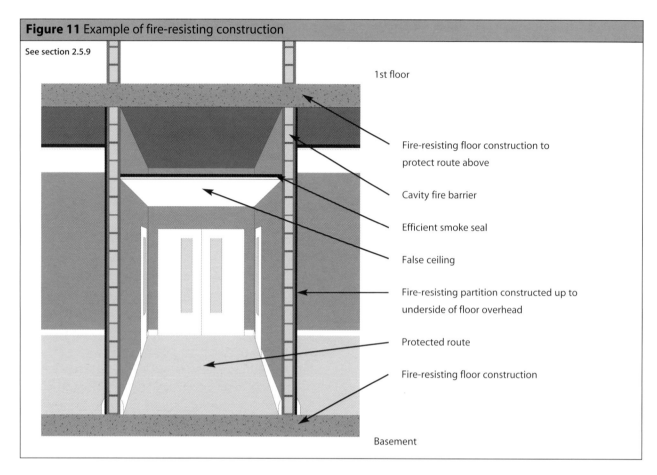

Figure 11 Example of fire-resisting construction

See section 2.5.9

1st floor

Fire-resisting floor construction to protect route above

Cavity fire barrier

Efficient smoke seal

False ceiling

Fire-resisting partition constructed up to underside of floor overhead

Protected route

Fire-resisting floor construction

Basement

2.5.9 Restricting the spread of fire and smoke through the school

Fire-resisting and smoke restricting construction has three primary objectives:

- to prevent fire and smoke from spreading into protected routes, ie, protected corridors and stairways;

- to isolate areas where the risk assessment has identified hazardous areas or areas identified as critical to the functioning of the school; and

- to restrict disproportionate damage to the school as a result of a fire by means of compartmentation thus limiting the fire to the place of origin.

Most of the detailed design guidance in later sections is concerned to a greater or lesser degree with restricting the spread of fire and smoke. Refer to sections 4 – 7 and appendix A in particular.

Fire resistance may defined in terms of three components: structural (loadbearing) capability (R), integrity (E) or insulation (I). See section 6.1.3.1 and appendix A for more details.

Compartmentation does not only serve as a barrier to fire and smoke spread, although that is the primary function. Compartment sizes also provide a degree of control over the fire load present in a compartment (which tends to be a function of room use and floor area) and hence the fire severity.

2.5.10 Fire doors

Any door in a fire-resisting or compartment wall will be a fire door, designed to resist the passage of fire and smoke (when closed). Fire doors are used on escape routes to sub-divide long corridors and thus ensure that no more than a short stretch of corridor leading to an exit is likely to become smoke-logged during a fire. Similarly, fire doors are used to separate stairs from circulation routes in order to protect the stairs from smoke ingress.

Fire doors will generally have glass vision panels to assist all occupants, including those with special needs, in their movement around the building. Vision panels will also be valuable in the event of fire, to enable people to see whether the space on the other side is affected by smoke or not. The extent of glazing that may be used depends on whether or not it provides insulation as well as integrity to the appropriate fire resistance level.

Hold open devices on fire-resisting self-closing doors on circulation routes will allow staff and pupils to move through the building freely. The hold open devices can be integral within the door closing device attached to the door or stand alone. In either case they must be connected to the automatic fire detection and alarm (AFD) system so the devices will release immediately when the detectors are actuated and the doors will close. This will ensure that compartmentation is maintained.

Pointers to the use of fire doors, as part of the detailed design guidance, are given in section 3.5. Details of the performance requirements for fire doors can be found in appendix C.

2.6   Arson

Good security is a major factor in reducing the incidence of arson fires. Measures to control access to the school buildings will need to be carefully planned in conjunction with other agencies such as the police and fire service.

Many arson attacks will be opportunistic and therefore whilst the removal of likely fuel, rubbish, etc, will reduce the possibility of an ignition occurring, even the best design will not counter the efforts of the really determined arsonist. But there are several strategies for making their efforts less effective.

For background information to the design against arson see the guide *How to combat arson in schools, July 1998,* from the Arson Prevention Bureau. The guide discusses the nature of arsonists and gives guidance on assessing a school's vulnerability to arson attack. It also contains detailed guidance on developing an action plan against arson.

To reduce arson the main objectives should be to:

- deter unauthorised entry onto the site;
- prevent unauthorised entry into the buildings;
- reduce the opportunity for an offender to start a fire;
- reduce the scope for potential fire damage; and
- reduce subsequent losses and disruption resulting from a fire.

The guide contains specific advice on how the five main objectives might be met and this guidance can be applied to the design of new schools or the extension/refurbishment of existing schools. The police and fire authority should be consulted on any measures considered necessary to reduce the threat of arson on a continuing basis during the life of the school; property insurers will also be able to advise on risk reduction measures.

2.6.1   Deterring unauthorised entry onto the site

The form of the boundary security should be chosen to suit the location of the school and the level of risk. Access to the site should ideally be controlled in some manner, and access to roofs from surrounding buildings or walls prevented.

2.6.2   Preventing unauthorised entry into a building

Door and frame construction and installation, with security locks and robust door and window ironmongery, will be beneficial in keeping intruders at bay.

There should be a limited number of entrances to the site and buildings. In particular, the number of concealed entrances or areas which offer cover to intruders should be kept to a minimum. Landscaping may be employed to enhance the building without providing cover to intruders.

Access to roofs (eg, from surrounding buildings or walls, or from within the site) should be prevented.

The outside of the premises should be well lit, particularly in the case of hidden or vulnerable areas.

The use of CCTV surveillance may act as a deterrent, although some insurers only believe it is beneficial if constantly monitored enabling a rapid response by security staff. Similar comments apply to the use of security alarms.

2.6.3 Reducing the opportunity for an offender to start a fire

Rubbish containers should be stored in locked areas well away from the buildings, school entrances and exits.

Combustible materials should be kept in secure zones, rooms, stores, etc, within the building.

All storerooms, staffrooms, the head teachers office and general office areas should be secured against intrusion at the end of the working day.

Combustible materials or rubbish should be kept in secure locked stores away from buildings or boundaries.

2.6.4 Reducing the scope for potential fire damage

To reduce the effects of arson, installation of a suppression system (eg, sprinklers) is likely to be very effective.

Automatic fire detection and alarm systems may be effective when the building is occupied, if provision is also made for first-aid fire-fighting (eg, extinguishers).

When common areas of the premises are unoccupied, linking an automatic fire detection system to an alarm receiving centre will enable the Fire and Rescue Service to make the earliest possible response to a fire.

Cloakrooms are areas where arson attacks occur relatively frequently. They should be treated as places of special fire hazard (see sections 3.1.2 and 6.3.2.2).

The construction of the external envelope of school buildings can reduce their vulnerability to arson attack. Unauthorised entry and the risks of arson damage by burning materials thrown into the building may be inhibited by the use of standard laminated security glass in windows and external doors. Such measures may be particularly employed for areas of higher risk, such as computer rooms, laboratories and storage areas. It should be noted, however, that such standard security glass laminates do not have significant fire resistance.

Waste bins should not be wall-mounted beneath windows and beneath combustible eaves.

2.6.5 Reducing subsequent losses and disruption resulting from a fire

Losses and disruption will be minimised if the fire can be kept small/contained, hence the measures in the preceding section will be beneficial here too.

Containment of the fire to its place of origin can be effectively encouraged by sound and solid building design and construction using the principles of compartmentation (see section 2.5.9). Sprinklers also function optimally when used in conjunction with a foundation for fire protection based on a compartmented and robust construction.

Valuable equipment should be protected by storing it within fire-resisting areas/containers. Computer files should be backed up and copies of key records kept (not in the same location as the originals.

It is also advisable to have a recovery plan in the event of fire (not just for arson).

Fire safety management issues are covered in more depth in the DCSF Managing School Facilities guides (see section 2.6.7).

2.6.6   Security, and prevention of arson

Fire safety including arson prevention is now closely tied to security of the buildings and the site. Good security along with good housekeeping will mean there is very little chance of the opportunist fire setter causing damage.

Schools are targets for petty theft and vandalism as well as deliberate fire setting. The standards outlined in the box below are designed to prevent intruders gaining access to the building and so will also be useful in reducing arson. These standards are recognised by Association of Chief Police Officers (ACPO) and the Association of British Insurers (ABI).

Examples of security standards

LPS1175 – *Doors, windows, shutters other façade elements and enclosures.*

LPS1214 – *IT theft*. This should also help with deterring entry by those who target schools to steal IT equipment and then set fire to the school to destroy evidence.

LPS1602/1603 – *Intruder Detection and Alarm systems*. These standards ensure that an intruder system is compliant with current legislation and are aimed at reducing false alarms.

The standards referred to above are intended to help with keeping out and/or detecting an intruder and by making it very difficult to gain entry to rooms or to steal equipment if already within the school.

The documents referred to above provide guidance on the test procedures for building elements with a security function. They are specific to the requirements of the proprietary LPCB certification schemes and may not necessarily be representative of requirements for compliance with other similar certification schemes for security products.

2.6.7   Further information on arson prevention

The Department for Children, Schools and Families has published Managing School Facilities Series Guides on Security (No 4) and Fire Safety (No 6) which taken together will eliminate many incidents as access to the buildings will be controlled with good security. More recently the Department has published *Key design guidance for schools, 2004*, which addresses the issue of security in schools with particular reference to the *Secured by Design (SBD)* (ref) approach backed by ACPO.

There is a DCSF online assessment for schools to find out if the are at risk from arson at www.teachernet.gov.uk/emergencies/resources/arson/index.html

The Home Office Crime Reduction website offers guidance in the form of a toolkit, see www.crimereduction.gov.uk/toolkits/ssh00.htm for details.

Further guidance on the reducing the risk of arson has been published by the Arson Prevention Bureau – see the website www.arsonpreventionbureau.org.uk for more information.

2.7   Designing for fire safety management

Management of fire safety must be integrated with all other management systems in the school. If this management is lacking then there is a danger that all the other areas such as security measures and alarm systems will be ineffective. To ensure there is no doubt as to where the responsibility for fire safety rests, and to enable consistency of approach, it is important that each establishment appoints a designated Fire Safety Manager. This should be a senior appointment preferably at Head or Deputy-Head level. It may be possible to appoint a professional to take on this role but that will depend on the size of the premises, costs, etc.

The designer should seek to understand how the building will be managed and seek to design the building and provide safety systems which can be readily and effectively managed and maintained.

The appointed person should have the necessary authority and powers of sanction to ensure that standards of fire safety are maintained.

The main duties of the Fire Safety Manager include:

- managing the school to minimise the incidence of fire (fire prevention); eg, good housekeeping and security;

- producing an Emergency Fire Plan;

- checking the adequacy of fire-fighting equipment and ensuring its regular maintenance;

- ensuring fire escape routes and fire exit doors/passageways are kept unobstructed and doors operate correctly;

- ensuring that fire detection and protection systems are maintained and tested and proper records are kept; and

- ensuring any close down procedures are followed.

The designer needs to take account of these duties, and make it as easy as possible for the Fire Safety Manager to carry them out.

There is considerable guidance on how to meet the above duties, and others, contained in *Fire Safety Guide No.6* (DfEE) in the *Managing School Facilities* series and in the *Educational Premises* guide in support of the Regulatory Reform (Fire Safety) Order 2005. There is also guidance in BS 5588: Part 12 *Fire Safety Management in buildings*.

Given the likelihood of school premises being used for other activities outside normal school hours, it is important that the designer includes this aspect when considering the fire safety

issues. There is a cost to implementing these actions and the school will be expected to implement those that are reasonable and within their financial limits. Anything beyond will probably need to be discussed with the LEA or building owner to obtain additional funding.

BB70 *Maintenance and renewal in educational buildings – maintenance of mechanical services* offers practical guidance. It points out that carrying out maintenance on the basis of emergency need may cost a lot more than pre-planned maintenance. The economic costs of components are covered as well as their life expectancy.

Thus a longer term view of the building and its needs in life cycle terms should be taken on board by the designer, rather than just handing over a finished building or extension. The Department for Children, Schools and Families has issued guidance on life cycle costing available on its AMP website www.teachernet.gov.uk/amps

The perception of the building as possessing a life cycle from green field to brown field is inherent in the thinking for fire safety engineering and the constant updating of the needs of the building at the various stages.

Provided the above approach to the design is adopted then the everyday management can be easily dealt with by the school. Escape routes need to be kept free of clutter and the fire behaviour of the controlled linings should not become compromised. Whilst this is primarily a management issue it is recommended that as circulation spaces are ideal for dispensing information the designer may wish to put display material on the escape route which can be behind a transparent cover to hold it in place. To reduce the effects of vandalism the cover should be of a material such as polycarbonate rather than glass as it will not shatter and is extremely difficult to ignite.

Section 3
Detailed design guidance

This section contains two main topics:

a. discussion of some issues connected with various types of rooms that are considered to be places of special fire hazard; and

b. sections dealing with various fire protection systems, namely:

- fire detection and alarm systems;
- sprinkler systems;
- smoke control systems; and
- fire doors,

indicating the various benefits that these systems may provide in accordance with the guidance enabling the Building Regulations to be satisfied.

Designers may wish to exploit the possibilities of using various fire protection systems and strategies to provide an alternative solution to that given in the detailed design guidance. In this case, the Building Control Body and/or the Fire and Rescue Service should be consulted at an early stage of the project, and the procedures outlined in section 2.3 (Fire safety engineering) should be followed.

The bulk of the detailed design guidance is contained in the following sections (4 – 8) and the appendices of this document.

3.1　Places of special fire hazard

This section deals with areas needing special consideration as they may either be High Hazard, ie, the source of 'hot' activities thus contributing to a high risk area, or may represent a valuable resource that is difficult to replace.

Places of special fire hazard that require additional protection in order to satisfy life safety requirements include the following:

- boiler rooms;
- storage space for fuel or other highly flammable substances (including PE mats) or chemicals;
- laboratories;
- technology rooms with open heat sources;
- kitchens;
- oil-filled transformer and switch-gear rooms; and
- rooms housing a fixed internal combustion engine.

Cloakrooms should also be regarded as places of special fire hazard (they are effectively storage spaces for highly flammable materials – hanging coats), by virtue of the relatively large number of arson fires that occur in them.

General guidance for the protection of such places is that they should be enclosed with fire-resisting construction, see section 6.3.2.2.

There are other specific areas that, whilst not falling into the category of special hazard, nevertheless would benefit from particular measures to enhance property protection. These areas include:

- ICT rooms;
- corridors and circulation spaces; and
- temporary and relocatable accommodation.

3.1.1　Storage spaces

Waste materials should be stored in wheelie bins in locked stores well away from the main buildings particularly if the materials are combustible; if they do not have a dedicated compound in an existing building, they should be secured to a wall away from the main buildings by a padlock and chain.

Storage areas within a building should be secure. Flammable sports mats should be stored in locked cupboards or stores.

While not a high hazard, the risk of losing irreplaceable items such as course work would be reduced if these were kept in fire protected storage.

3.1.2　Cloakrooms

Cloakrooms are high-risk areas due to the nature of the fire load (coats) and the relatively large number of daytime arson fires they experience. Cloakrooms should not form part of or be open to circulation spaces. If it is not feasible for them to be located within fire-resisting enclosures, then lockers made from materials of limited combustibility (as defined in appendix A) should be provided. If these lockers line a corridor, the restrictions of section 3.1.6 apply.

3.1.3　Laboratories and technology rooms

Science laboratories and preparation rooms, and some design technology areas, are commonly fitted with gas supplies. Advice is given in the

Institute of Gas Engineers 2004 publication *UP11 Gas installations for educational establishments* and IM/25 (ref). Each laboratory should be fitted with a lockable isolating valve to enable gas supplies to gas taps on benches to be shut off at the end of the day.

Where fume cupboards are provided these should have a powered extract system and fire-resisting ductwork to extract hot smoke in case a fire occurs within them.

In order to improve property protection it is suggested that all science departments are provided with a central fire-resisting storage cupboard for storage of flammable materials.

3.1.4 Kitchens

Consideration should be given to providing a proprietary extinguishing system in cooker hoods. This will be of benefit to both life safety and the minimisation of fire damage.

If gas supplies are fitted, refer to advice given in the Institute of Gas Engineers 2004 publication *UP11 Gas installations for educational establishments*. Where gas cookers are provided in classrooms (eg, for home economics/domestic science lessons), the room should be fitted with a lockable isolating valve to enable gas supplies to cookers to be shut off at the end of the day.

3.1.5 ICT areas

Valuable, specialist equipment may be susceptible to damage by heat or smoke as well as the means of extinguishing the fire, eg, water. There is usually a low risk of fire but rooms that such equipment is stored in can be enclosed in fire-resisting construction to protect them from fire spread from elsewhere in the building.

3.1.6 Corridors and circulation areas

The schools' circulation routes will almost certainly be important for relaying information to the pupils by means of notice boards, or used as a display area for eg, pupils' work. Notice boards should not be more than 3m wide, and there should be a gap between notice boards on the same wall of at least 1m. Notice boards in a protected corridor should be fitted with a cover, preferably top hung so that the cover cannot be left 'jutting-out' into the escape route.

If a corridor is lined with lockers then these should be made from materials of limited combustibility (as defined in appendix A), and any rooms off the corridor should be regarded as inner rooms with the corridor treated as the access room (see section 4.3.2.6).

3.1.7 Temporary and relocatable accommodation

Temporary accommodation, intended to remain in place for less than 28 days, is not subject to the Building Regulations. Nevertheless, for fire safety, temporary and relocatable accommodation should be treated in the same way as a permanent structure.

The siting of temporary classrooms, etc, will need to be discussed with the fire authority so access for fire-fighting is available. Treat as any other building with regard to space separation (section 7.3), using the 'notional boundary' concept to set the appropriate distances for property protection.

To prevent combustible material accumulating beneath these rooms, it is advisable to consider fitting 'skirts' (eg, boarding material) around the bases of temporary/relocatable buildings.

3.1.8 Boiler rooms

School boiler/plant rooms are considered as areas of high fire risk. For this reason the following fire safety precautions are required for new schools and where an existing boiler/plant room is upgraded or refurbished:

- A means of automatically shutting off the fuel supply in the event of a fire. This should include an emergency shut-off push-button at the entrance to the boiler room. The system should shut off the electrical power to the plant. In the event of a genuine alarm, the system should require manual resetting, but if it is purely a power failure and appropriate self-proving devices are in place, then automatic resetting is appropriate to prevent the risk of pipe freezing during weekends or holiday periods. Alternatively a system of alarm notification to remote keyholders can be used. See BS6644 clause 6.8 *Additional safety controls* (**Note:** Many existing boiler rooms are fitted with a manual isolating valve on the fuel supply hence the requirement for automatic isolation of the gas supply to new buildings is not intended to be retrospectively applied to all existing buildings).

- Heat detection linked to the fire alarm system to raise the alarm. Heat detection is preferable to smoke detection in a boiler room as smoke detection is more likely to cause false alarms.

- For boiler rooms with difficult access or located inside or connected to a building, a foam inlet point and sometimes smoke vents are often required. With large installations (200kW and above) and especially with oil tanks within the building, foam spray heads are required with a pipe to outside for connection by the Fire and Rescue Service.

In addition, a carbon monoxide or carbon dioxide detection system is often required in boiler/plant rooms for all fuel types, particularly where a boiler/plant room forms part of the school building itself. The detection system should both raise an alarm and isolate the fuel supply.

For gas fired boilers it is recommended that the detection system is combined with an unburned gas detection system that is appropriately located within the boiler/plant room depending on the fuel type (natural gas at high level and Liquefied Petroleum Gas at low level). Again, the detection system should both raise an alarm and isolate the fuel supply.

References:

Gas – BS 6644: 2005 *Specification for the installation of gas-fired hot water boilers of between 70kW (net) and 1.8MW (net) (2nd and 3rd family gases)*, IGE/UP/10 Edition 3 *Installation of flued gas appliances in industrial and commercial premises*, IGE/UP/11 *Gas in educational establishments*.

Oil – BS 5410 Parts 1 & 2, *Oil Firing* and BS799 Part 5, *Oil Burning Equipment*.

3.2 Fire detection and alarm systems

All schools must have a means of raising the alarm, but there are many factors that affect the choice of detection system.

- Is it for life safety and/or property protection? ie, an L system (section 4.2) and/or a P system (section 4.2.1.5)?

- Should it be an automatic alarm system or an M system with manual call points triggered by occupants (section 4.2.1.1)?

- Will it need to be linked to a central monitoring station (section 4.2.1.5)?

- Will it need to be linked to hold-open devices on doors (section 4.2.1.4)?

- What sort of sounders are required?

- Should there be a voice alarm system tailored to the needs of the school (section 4.2)?

- Are any of the people present likely to be hearing impaired? What means are there to alert them to a fire (section 4.2.1.2)?

Guidance in respect of fire detection and alarm systems is set out in section 4.2.

Automatic systems may provide a number of benefits. They can be used to control other systems such as smoke control (section 3.4 and 4.2.1.4), automatic fire suppression systems, hold-open devices on fire doors, etc (see section 4.2.1.4), and shut down mechanical ventilation systems (section 4.5.7).

Property protection can be enhanced if the alarm signal has a direct link via an Alarm Receiving Centre (ARC) to initiate a response by the Fire and Rescue Service.

Some less obvious benefits are:

- They are one of the possible options that allow 'inner rooms' to be used (see section 4.3.2.6).

- In some cases cavities of 'unlimited' size (ie, without cavity barriers) may be permitted (section 6.4.4.1).

3.3 Sprinkler systems

Sprinkler systems, by controlling or extinguishing fires while they are still small, have obvious benefits for reducing fire damage. The sprinkler systems can be designed to sound an alarm via an Alarm Receiving Centre (ARC) to initiate a response by the Fire and Rescue Service (section 4.2.1.5) which is a further benefit for property protection.

If the fire is kept small or put out, this will also significantly reduce the amount of smoke produced, which buys more time for evacuation before escape routes become untenable due to smoke logging. Despite this, the specific areas in the detailed design guidance where 'life safety' sprinkler systems are given as an alternative option are actually quite limited.

These cases are:

- Compartment sizes may be considerably larger if the building is fitted with sprinklers (table 9, see section 6.3.2).

- In building separation distance calculations, the assumption is made that sprinklers will moderate the fire intensity (section 7.3.1). The result is that the boundary distance halves if the building has sprinklers, or equivalently, the unprotected area may be doubled (see section 7.3.4.2). Sprinklers may be accounted for in similar way if using BRE calculations instead (as per section 7.3.4).

- If the building has a footprint of greater than 900m² on any storey where the floor is more than 7.5m above the level of Fire and Rescue Service access, fire-fighting shafts and fire mains will be required (see section 8.4.3). If the building has sprinklers, the fire-fighting shafts may be up to 60m from a fire main outlet, rather than 45m (section 8.4.3). In other words, fewer shafts may be required to give access to the whole of the floor area.

- Basements may have mechanical extraction of smoke, rather than provide a vent to the outside, if the basement has sprinkler protection (see section 8.5.2.2).

- The minimum period of fire resistance for a low-rise building (top floor less than 5m above ground level) is 30min, rather than 60min, if the building is fitted with sprinklers (see appendix A, table A2).

It is worth emphasising that these cases **only** apply if the sprinkler system qualifies as a 'life safety' system. The distinction between life safety and property protection systems is explained in the SSLD document on sprinklers in schools (ref).

For the purposes of property protection, the detailed design guidance does not require all floors to be compartment floors, if the building is protected with sprinklers (section 6.3.2.1). In this specific case, the system does not need to be a life safety system.

Briefly, both types of sprinkler systems provide the same protection of life in terms of their ability to control and suppress a fire. However, life safety systems have additional features intended to

increase the system reliability and on-going availability.

After performing the risk assessment (section 2.4.1) it is anticipated that a property protection system will be recommended for all but the lowest-risk schools. Life safety systems will be required when life safety fire protection or fire safety measures are dependent on sprinklers, eg, either to meet Building Regulations or RRO requirements, or as an alternative solution.

For further guidance on the benefits of sprinklers, the designer can refer to *Sprinklers for safety: the use and benefits of incorporating sprinklers in buildings and structures*, BAFSA 2006.

3.4 Smoke control systems

The detailed design guidance in sections 4 – 8 mainly relies on containment by walls and doors to prevent smoke from entering escape routes. Automatic smoke control systems (triggered by an automatic detection system) can achieve the same ends, by extracting smoke from the building. This provides benefits for means of escape and also facilitates access by the Fire and Rescue Service. As with sprinkler systems, though, the cases in the detailed guidance in which automatic smoke control systems are given as an alternative option are limited.

These cases are:

- There is a requirement for the venting of smoke from basements – smoke control may be used if a sprinkler system is installed (section 8.5.2.2).

- In making adequate provision for means of escape, it is not necessary to discount a stair if it is protected, either by a lobby (FR construction) or by smoke control (section 4.4.5.2).

- In some cases it may be acceptable for a single stair to serve the building (or part of the building) (section 4.4.6.3) if it has lobby protection or smoke control.

- If stairs are pressurised to prevent smoke ingress, the doors leading to the stairs may have less stringent requirements with regard to smoke leakage (appendix C, table C1, note 2a).

- If the building requires fire-fighting shafts (section 8.4.2) then these should have a smoke control system (figure 46, note 2).

The free area of extraction vents is defined in appendix E (figure E6).

Mechanical ventilation and air conditioning systems are covered by section 4.5.7. It is worth considering whether such systems could be designed in such a manner as to provide a smoke control function during a fire.

3.5   Fire doors

Within the detailed design guidance, fire doors are the normal method for preventing fire/smoke spread into/within escape routes. Openings in compartment walls (see sections 6.3.4.1 – 6.3.4.3), protected shafts (see section 6.3.5.6) and fire-resisting construction will need to be fire doors. Other circumstances where fire doors are used are in the sub-division of corridors (see section 4.3.2.16), and between stairs and circulation routes (see section 4.3.2.10), to prevent smoke-logging of the stairs.

One area in which fire doors may be considered as an option is the protection of stairs by a fire-resisting lobby. In such cases the likelihood of a stair not being available is significantly reduced and it is not necessary to discount a stair in order to ensure that the capacity of the remaining stair(s) is adequate for the number of persons needing to escape (see section 4.4.5.2).

Further detailed guidance in respect of fire doors is set out in appendix C.

3.6   Methods of measurement

Some form of measurement is an integral part of much of the guidance in this document and methods are set out in appendix E.

3.7   Fire performance of materials, products and structures

Much of the guidance throughout this document is given in terms of performance in relation to standard fire test methods. Details are drawn together in appendix A to which reference is made where appropriate. In the case of fire protection systems, reference is made to standards for systems design and installation.

Standards referred to can be found on the British Standards Institute (BSI) website.

3.8   Robust construction

By the nature of the activities that go on within a school and the exuberance of youth, the elements of construction making up a school should be able to withstand more abuse than those incorporated in the normal environment. This is particularly true in areas within the building that are used for physical training, sports or play.

Similarly, many modern products used in the construction of fire-resisting barriers are sensitive to both installation, maintenance and careful use in practice. Non-robust and use-sensitive products should only be used with caution.

The following references may be found helpful:

- All glass should meet the requirements of BS 6262-4, BS 6180 (if used in a barrier) and BS 5234 (if used in a partition).

- All glass used in the construction of fire walls in schools should be classed B or C in accordance with BS6206: 1981 dependent on location and pane size.

- All fire separating walls should, as well as providing the rating specified above, be rated at least heavy duty (HD) when evaluated by BS 5234: Part 2; 1992 *Specification for performance requirements for strength and robustness*, and should be installed in accordance with Part 1 of BS 5234.

- Fire-resisting timber door assemblies should achieve a performance rating of Severe Duty (SD) in accordance with DD 171: 1987.

- Door hardware will be essential or non-essential depending on door function and fittings, see BS 8214: 1990.

Third party certification schemes for fire protection products and related services are an effective means of providing assurances about the level of quality, reliability and safety that non-certificated products may lack. While this does not mean that goods that are not third party approved are less reliable, there is no obvious way of demonstrating expected performance.

Section 4
Means of warning and escape

4.1   Overview

4.1.1   Requirement B1 of the Building Regulations

The building shall be designed and constructed so that there are appropriate provisions for the early warning of fire, and appropriate means of escape in case of fire from the building to a place of safety outside the building capable of being safely and effectively used at all material times.

4.1.2   Performance

The Requirement B1 will be met if:

- there are routes of sufficient number and capacity, which are suitably located to enable persons to escape to a place of safety in the event of fire;

- the routes are sufficiently protected from the effects of fire where necessary;

- the routes are adequately lit;

- the exits are suitably signed;

- there are appropriate facilities to either limit the ingress of smoke to the escape route(s) or to restrict the fire and remove smoke;

all to an extent necessary that is dependent on the use of the building, its size and height, and

- there is sufficient means for giving early warning of fire for persons in the building.

4.1.3   Introduction

In an emergency such as a fire, all the occupants should be able to reach a place of safety without delay. There shall be sufficient exit routes and doors to allow everyone to get to the final exit and then away from the building.

The design of the escape routes will depend on the complexity of the layout of the building, how large it is, how many floors there are and the mobility of the occupants. Where possible all escape routes should mirror the everyday circulation patterns within the building. This avoids providing alternative means of escape that are only used in an emergency. This will also ensure that all the pupils, especially the youngest ones, are already familiar with how to leave the school quickly and will minimise their anxiety in an emergency particularly while the fire alarms are sounding.

School buildings are often used by large numbers of people other than pupils and staff who may be unfamiliar with the layout and account should be taken of this intended use.

These provisions relate to building work and material changes of use which are subject to the functional requirement B1 and they may therefore affect new or existing buildings. They are concerned with the measures necessary to ensure reasonable facilities for means of escape in case of fire. They are only concerned with structural fire precautions where these are necessary to safeguard escape routes.

They assume that, in the design of the building, reliance should not be placed on external rescue by the Fire and Rescue Service nor should it be based on a presumption that the Fire and Rescue Service will attend an incident within a given time. This guidance has been prepared on the basis that, in an emergency, the occupants of any part of a building should be able to escape safely without any external assistance.

4.1.3.1   Analysis of the problem

The design of means of escape and the provision of other fire safety measures such as a fire alarm system (where appropriate), should be based on an assessment of the risk to the occupants should a fire occur. The assessment should take into account the nature of the building structure, the use of the building, the processes undertaken and/or materials stored in the building; the potential sources of fire; the potential of fire spread through the building; and the standard of fire safety management proposed. Where it is not possible to identify with any certainty any of these elements, a judgement as to the likely level of provision must be made.

Fires do not normally start in two different places in a building at the same time. Initially a fire will create a hazard only in the part in which it starts and it is unlikely, at this stage, to involve a large area. The fire may subsequently spread to other parts of the building, usually along the circulation routes. The items that are the first to be ignited are often furnishings and other items not controlled by the regulations. It is less likely that the fire will originate in the structure of the building itself and the risk of it originating accidentally in circulation areas, such as corridors, lobbies or stairways, is limited, provided that the combustible content of such areas is restricted.

The primary danger associated with fire in its early stages is not flame but the smoke and noxious gases produced by the fire. They cause most of the casualties and may also obscure the way to escape routes and exits. Measures designed to provide safe means of escape must therefore provide appropriate arrangements to limit the rapid spread of smoke and fumes.

Protection of property, however, largely aims to mitigate the development of the fire after flashover when flames, heat and hot gases will present the most danger to the building fabric. This will also provide added life safety protection for any occupants that remain trapped within the building or fire fighters who have cause to enter the building.

4.1.3.2 Criteria for means of escape

The basic principles for the design of means of escape are:

- that there should be alternative means of escape from most situations; and

- where direct escape to a place of safety is not possible, it should be possible to reach a place of relative safety, such as a protected stairway, which is on a route to an exit, within a reasonable travel distance. In such cases the means of escape will consist of two parts, the first being unprotected in accommodation and circulation areas and the second in protected stairways (and in some circumstances protected corridors).

Note: Some people, for example those who use wheelchairs, may not be able to use stairways without assistance. For them evacuation involving the use of refuges on escape routes and either assistance down (or up) stairways or the use of suitable lifts will be necessary.

The ultimate place of safety is the open air clear of the effects of the fire.

The following are not acceptable as means of escape:

- lifts (except for a suitably designed and installed evacuation lift – see section 4.5.6.1);

- portable ladders and throw-out ladders;

- manipulative apparatus and appliances, eg, fold-down ladders and chutes; or

- helical/spiral stairs (not suitable for pupils or members of the public).

Escalators should not be counted as providing predictable exit capacity, although it is recognised that they are likely to be used by people who are escaping. Mechanised walkways could be accepted and their capacity assessed on the basis of their use as a walking route, while in the static mode.

4.1.3.3 Alternative means of escape

There is always the possibility of the path of a single escape route being rendered impassable by fire, smoke or fumes. Ideally, therefore people should be able to turn their backs on a fire wherever it occurs and travel away from it to a final exit or protected escape route leading to a place of safety. However, in certain conditions a single direction of escape (a dead end) can be accepted as providing reasonable safety. These conditions depend on the use of the building and its associated fire risk, the size and height of the building, the extent of the dead end and the numbers of persons accommodated within the dead end.

4.1.3.4 Unprotected and protected escape routes

The unprotected part of an escape route is that part which a person has to traverse before reaching either the safety of a final exit or the comparative safety of a protected escape route, ie, a protected corridor or protected stairway.

Unprotected escape routes should be limited in extent so that people do not have to travel excessive distances while exposed to the immediate danger of fire and smoke. Even with protected horizontal escape routes, the distance to a final exit or protected stairway needs to be limited because the structure does not give protection indefinitely.

Protected stairways are designed to provide virtually 'fire sterile' areas which lead to places of safety outside the building. Once inside a protected stairway, a person can be considered to be safe from immediate danger from flame and smoke. They can then proceed to a place of safety at their own pace. To enable this to be done, flames, smoke and gases must be excluded from these escape routes, as far as is reasonably possible, by fire-resisting structures. This does not preclude the use of unprotected stairs for day-to-day circulation, but they can only play a very limited role in terms of means of escape due to their vulnerability in fire situations.

4.1.3.5   Security

The need for easy and rapid evacuation of a building in case of fire may conflict with the control of entry and exit in the interest of security. Measures intended to prevent unauthorised access can also hinder entry of the Fire and Rescue Service to rescue people trapped by fire.

Potential conflicts should be identified and resolved at the design stage and not left to ad hoc expedients after completion. The architectural liaison officers attached to most police forces are a valuable source of advice. Some more detailed guidance on door security in buildings is given in section 4.5.3.

4.2   Fire alarm and fire detection systems

All schools should have arrangements for raising the alarm in the event of a fire. In existing small schools on one storey, with no more than 160 pupils, the means of raising the alarm may be simple. For instance, manually operated sounders (such as rotary gongs or hand bells) may be used. However, it must be determined that the warning can be heard and understood throughout the premises, including for example the toilet areas.

In all other cases, the school should be provided with a suitable electrically operated fire warning system in accordance with BS 5839-1:2002 *Fire detection and alarm systems for buildings, Code of practice for system design, installation commissioning and maintenance.*

BS 5839-1 specifies three categories of system, ie, category 'M' for manual alarm systems; category 'L' for the protection of life; and category 'P' for property protection (see section 4.2.1.5).

The fire alarm may be used as a class change signal in schools to indicate start or finish of pre-determined periods. To avoid the risk of confusion the duration of such class change signals should not exceed five seconds. This dual use can be difficult to arrange as the sound levels that are acceptable for a fire alarm are too high for class change and levels suitable for class change will result in more sounders than for a dedicated fire alarm sounder system.

Consideration should be given to installing a voice alarm system. Such a system could form part of a public address system and give both an audible signal and verbal instructions in the event of fire. The fire warning signal should be distinct from other signals which may be in general use and be accompanied by clear verbal instructions. If a voice alarm system is to be installed, it should comply with BS 5839-8:1998 *Code of practice for the design, installation and servicing of voice alarm systems.*

In general, a category M system will satisfy the Building Regulations and other statutory requirements for schools. However, there are often circumstances where a category L fire detection system may be needed. For example:

- to compensate for some departure from the guidance elsewhere in this document;

- as part of the operating system for some fire protection systems, such as pressure differential systems or automatic door releases;

- where a fire could break out in an unoccupied part of the premises (eg, a storage area or basement that is not visited on a regular basis, or a part of the school that has been temporarily vacated) and prejudice the means of escape from any occupied part(s) of the school (L2); and

- where the school is likely to be partially occupied outside normal school hours (L4).

Category L systems are sub-divided into:

L1 – systems installed throughout the protected building;

L2 – systems installed only in defined parts of the protected building (a category L2 system should normally include the coverage required of a category L3 system);

L3 – systems designed to give a warning of fire at an early enough stage to enable all occupants, other than possibly those in the room of fire origin, to escape safely, before the escape routes are impassable owing to the presence of fire, smoke or toxic gases;

L4 – systems installed within those parts of the escape routes comprising circulation areas and circulation spaces, such as corridors and stairways; and

L5 – systems in which the protected area(s) and/or the location of detectors is designed to satisfy a specific fire safety objective (other than that of a category L1, L2, L3 or L4 system).

4.2.1.1   Manual Call Points

Call points for electrical alarm systems should comply with BS 5839-2:1983, or Type A of BS EN 54-11:2001 and these should be installed in accordance with BS 5839-1. Type B call points should only be used with the approval of the Building Control Body.

BS EN 54-11 covers two types of call points:

Type A (direct operation) in which the change to the alarm condition is automatic (ie, without the need for further manual action) when the frangible element is broken or displaced; and

Type B (indirect operation) in which the change to the alarm condition requires a separate manual operation of the operating element by the user after the frangible element is broken or displaced.

Wherever possible manual call points should be located such that they are less likely to be prone to malicious operation. This can sometimes be achieved by ensuring that they are in open view of staff. Where manual call points are still likely to be subject to casual, malicious operation, it is acceptable for a transparent, hinged cover to be fitted to the manual call points. Operation of this two-action manual call point then involves

lifting the cover and operating the manual call point in the normal manner.

Manual call points should be located so that the travel distance from any part of the building to the nearest one is not more than 45m. The recommended height above the floor is 1.4m. There should be at least one call point per floor. There should also be a call point in or near places of high hazard, and in the assembly hall.

4.2.1.2   Warnings for people with impaired hearing

A suitable method of warning (eg, a visual and audible fire alarm signal) should be provided in schools where it is anticipated that one or more persons with impaired hearing may be in relative isolation and where there is no other suitable method of alerting them.

In many cases a vibrating paging system may be more appropriate than fixed beacons. This could also be used for alerting people with other disabilities. Clause 18 of BS 5839-1:2002 provides detailed guidance on the design and selection of fire alarm warnings for people with impaired hearing.

4.2.1.3   Design and installation of systems

It is essential that fire detection and fire warning systems are properly designed, installed and maintained. Where a fire alarm system is installed, an installation and commissioning certificate should be provided. Third party certification schemes for fire protection products and related services are an effective means of providing the fullest possible assurances, offering a high level of quality, reliability and safety.

Alarm systems should be standardised across a school. However, systems in different buildings may be self-contained.

4.2.1.4   Interface between fire detection and fire alarm systems and other systems

Fire detection and fire alarm systems are sometimes used to initiate the operation, or change of state, of other systems, such as smoke control systems, fire extinguishing systems, release arrangements for electrically held-open fire doors and electronically locked exit doors. It is essential that the interface between the fire detection and fire alarm system and any other system required for compliance with the Building Regulations is designed to achieve a high degree of reliability. Particular care should be taken if the interface is facilitated via another system, such as an access control system. Where any part of BS 7273 applies to actuation of other systems, the recommendations of that standard should be followed. For example electronic access controls fitted to fire exit doors should fail in the open position not the closed position.

4.2.1.5   Property protection

For the purposes of property protection, arrangements for alerting the Fire and Rescue Service should be put in place so as to minimize the time between ignition and the arrival of fire-fighters. Any such arrangements should be such as to minimize the risk to any person responsible for summoning the fire service. Unless the school is continuously occupied at all times the guidance given in BS 5839-1 for category P systems should be adopted.

Type P systems are sub-divided into:

P1 – systems installed throughout the protected building; and

P2 – systems installed only in defined parts of the protected building.

Where a sprinkler system has been installed in the building then this can be linked to the alarm system as an alternative to providing fire detectors for property protection, however detectors may still be required for life safety.

The objective of property protection is unlikely to be satisfied, unless the system incorporates means for automatic transmission of alarm signals to the Fire and Rescue Service. Clause 15 of BS 5839-1 gives guidance on appropriate means of communicating with the fire service and provisions for the protection of communicating equipment. Any alarm receiving centre to which fire alarm signals are relayed should comply with the recommendations of BS 5979. Systems that automatically transmit a pre-recorded message direct via the public emergency call system should not be used.

4.3   Design for horizontal escape

4.3.1   Introduction

The general principle to be followed when designing facilities for means of escape is that any person confronted by an outbreak of fire within a building can turn away from it and make a safe escape.

Dead ends should not be included in new buildings. However, in certain conditions a single direction of escape (a dead end) can be accepted as providing reasonable safety. These conditions depend on the fire risk, the extent of the dead end and the numbers of persons accommodated within the dead end.

This section deals with the provision of means of escape from any point to the storey exit of the floor in question. It should be read in conjunction with the guidance on the vertical part of the escape route in section 4.4 and the general provisions in section 4.5.

4.3.2   Escape route design

4.3.2.1   Number of escape routes and exits

The number of escape routes and exits to be provided depends on the number of occupants in the room, tier or storey in question and the limits on travel distance to the nearest exit given in table 1.

Note: It is only the distance to the nearest exit that should be so limited. Any other exits may be further away than the distances in table 1.

In multi-storey buildings (see section 4.4) more than one stair may be needed for escape, in which case every part of each storey will need to have access to more than one stair. This does not prevent areas from being in a dead end condition provided that the alternative stair is accessible in case the first one is not usable.

In order to avoid occupants being trapped by fire or smoke, there should be alternative escape routes from all parts of the building. Dead end corridors should not be included in new buildings.

However, a single route is acceptable for:

a. parts of a floor from which a storey exit can be reached within the travel distance limit for travel in one direction set in table 1 (also see section 4.3.2.2). This is provided that no one room in this situation has an occupant capacity of more than 60 people. The calculation of occupant capacity is described in table 2; or

b. a storey with an occupant capacity of not more than 60 people, where the limits on travel in one direction only are satisfied (see table 1).

In many cases there will not be an alternative at the beginning of the route. For example, there may be only one exit from a room to a corridor, from which point escape is possible in two directions. This is acceptable provided that the overall distance to the nearest storey exit is within the limits for routes where there is an alternative and the 'one direction only' section of the route does not exceed the limit for travel where there is no alternative, see table 1. Figure 12 shows an example of a dead end condition in an open storey layout.

Very young children (infants/nursery school age) will move more slowly than older children or adults, and also require constant supervision and direction during egress. Consideration should be given to providing direct access to an external place of safety from their classrooms.

4.3.2.2   Access control measures

Measures incorporated into the design of a building to restrict access to the building or parts of it should not adversely affect fire safety provisions.

Whilst it may be reasonable to secure some escape routes outside normal business hours, the measures left in place should be sufficient to allow safe evacuation of any persons left inside the building (see section 4.5.3.1).

4.3.2.3   Limits on travel distance

Limiting the travel distance to a place of relative safety has a number of benefits, namely:

- limited travel time; safety may be reached without serious exposure to smoke;
- limited size and complexity of enclosure;
- provision of sufficient alternative escape capacity within a reasonable distance. If there is a choice of exits, occupants should be able to escape in a direction away from the fire;

- increased likelihood that an exit is visible, and remains so during a fire;
- reduced likelihood that a fire can occur unseen, or grow large before detection/alarm; and
- reduced likelihood of a fire between occupant and exit.

The maximum travel distance from any point within the building to a final exit or a storey exit is given in table 1 below.

Table 1 Guidance to suitable travel distances

Location	Maximum travel distance[1] where travel is possible in:	
	One direction only (m)	More than one direction (m)
Places of special fire hazard	9	18
Areas with seating in rows	15	32
Areas not listed above	18	45
Ground storey of small premises with a single exit	27	N/A

Notes:

1. The dimensions in the table are travel distances. If the internal layout of partitions, fittings, etc, is not known when plans are deposited, direct distances may be used for assessment. The direct distance is taken as 2/3rds of the travel distance.

Figure 12 Travel distance in dead end condition

See section 4.3.2.2

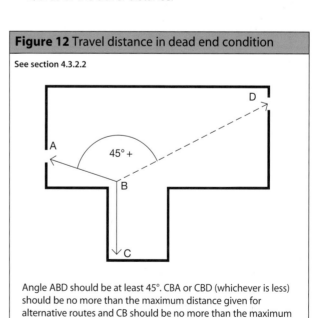

Angle ABD should be at least 45°. CBA or CBD (whichever is less) should be no more than the maximum distance given for alternative routes and CB should be no more than the maximum distance for travel where there are no alternative routes.

For existing buildings, the RRO Guide (ref) offers some flexibility in the suitable travel distances depending on whether the building or compartment has been assessed as low, average or high risk. The figures in table 1 correspond to 'average' risk; for low/high risk the distances lengthen/shorten respectively, by about 30% ~ 50%. Consult the RRO Guide Part 2, section 4 (ref) for more details.

4.3.2.4   Number of occupants and exits

The figure used for the number of occupants will normally be that specified as the basis for the design. When the number of occupants likely to use a room, tier or storey is not known, the capacity should be calculated on the basis of the appropriate floor space factors.

Table 2 Occupant capacity in rooms or areas	
Room/Area	Occupant capacity based on floor space factor (m²/person) or design intent
Classroom/Lecture Room/Study Room	Maximum design capacity (eg, no. of seats)
Dining Room	0.9
Assembly Hall/Dual Purpose Area	0.45
Sports Hall (not used for assembly or examinations, etc)	5.0
Store Room	30.0
Office	6.0
Staff Common Room	1.0

Table 3 gives the minimum number of escape routes and exits from a room or storey according to the number of occupants (this number is likely to be increased by the need to observe travel distances and by other practical considerations).

Table 3 Minimum number of escape routes and exits from a room, tier or storey	
Maximum number of persons	Minimum number of escape routes/exits
60	1
600	2
more than 600	3

The maximum number of persons given in table 3 is for 'normal' levels of risk. If the risk assessment shows that a room or area of an existing building is of 'low' or 'high' risk, then these numbers of people can be adjusted 'slightly', up or down respectively. Refer to the RRO guide Part 2, section 4 for more details.

4.3.2.5   Alternative escape routes

A choice of escape routes is of little value if they are all likely to be disabled simultaneously. Alternative escape routes should therefore satisfy the following criteria:

- they are in directions 45° or more apart (see figure 13); or
- they are in directions less than 45° apart, but are separated from each other by fire-resisting construction.

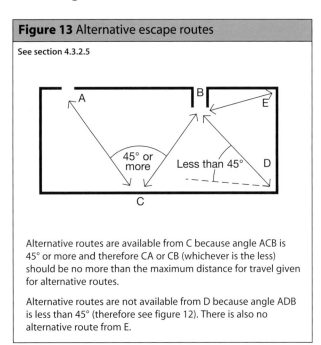

Figure 13 Alternative escape routes

See section 4.3.2.5

Alternative routes are available from C because angle ACB is 45° or more and therefore CA or CB (whichever is the less) should be no more than the maximum distance for travel given for alternative routes.

Alternative routes are not available from D because angle ADB is less than 45° (therefore see figure 12). There is also no alternative route from E.

Dead end corridors should be avoided in new construction.

In the case of an existing building, dead ends should conform with the travel distances given in table 1, and:

- automatic fire detection should be provided; or
- the escape route should have a fire resistance of at least 30 minutes.

Alternatively, an additional exit may be provided in order to eliminate the dead end condition.

4.3.2.6   Inner rooms

A room from which the only escape route is through another room is called an inner room. It is at risk if a fire starts in the other room, called the access room (see figure 14).

Such an arrangement is only acceptable if the following conditions are satisfied:

- the occupant capacity of the inner room should not exceed 60;

- the inner room should be entered directly off the access room (ie, there should not be a corridor between access room and inner room. See section 3.1.6 for circumstances in which a corridor or circulation space is regarded as an access room);

- the escape route from the inner room should not pass through more than one access room;

- the travel distance from any point in the inner room to the exit(s) from the access room should not exceed the appropriate limit given in table 1;

- the access room should not be a place of special fire hazard; and

- one of the following arrangements should be made:

  i. the enclosures (walls or partitions) of the inner room should be stopped at least 500mm below the ceiling; or

  ii. a suitably sited vision panel not less than 0.1m² should be located in the door or walls of the inner room, to enable occupants of the inner room to see if a fire has started in the outer room; or

  iii. the access room should be fitted with a suitable automatic fire detection and alarm system to warn the occupants of the inner room of the outbreak of a fire in the access room.

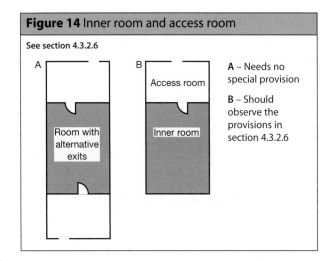

Figure 14 Inner room and access room

See section 4.3.2.6

A – Needs no special provision

B – Should observe the provisions in section 4.3.2.6

4.3.2.7   Planning of exits in a central core

Buildings with more than one exit in a central core should be planned so that storey exits are remote from one another and so that no two exits are approached from the same lift hall, common lobby or undivided corridor, or linked by any of these (see figure 15).

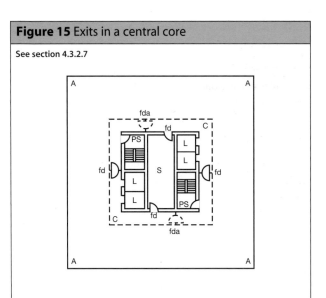

Figure 15 Exits in a central core

See section 4.3.2.7

Note: The doors at both ends of the area marked 'S' should be self-closing fire doors unless the area is sub-divided such that any fire in that area will not be able to prejudice both sections of corridor at the same time. If that area is a lift lobby, doors should be provided as shown in Figure 8 in BS 5588: Part 11: 1997.

Key:
L Lift
S Services, toilets, etc
fd Self-closing FD20S fire doors
fda Possible alternative position for fire door
C Corridor off which accommodation opens
PS Protected stairway
A Accommodation (eg, teaching space)

4.3.2.8 Open spatial planning

Where an open-plan space connects more than one storey, rooms accessed from the space should be treated as inner rooms (section 4.3.2.6) with the space/balcony regarded as the access room. Any escape routes should not be prejudiced by openings in floors.

Escape routes should not be within 4.5m of openings unless:

- the direction of travel is away from the opening (eg, A-B in figure 16a); or
- there is an alternative escape route which does not pass within 4.5m of the open connection (eg, the rooms with alternative exits in 16b).

In sprinklered schools, rooms which are accessed by an open balcony less than 4.5m wide, and which do not have an alternative escape route away from the balcony, should satisfy the following conditions (see figure 16b):

- escape from any point on the balcony should be available in at least two directions; and
- the travel distance along the balcony should not exceed 18m.

Note: If the opening passes through a compartment floor (see section 6.3.2.1) then the guidance in BS 5588 Part 7 (1997) should be followed for fire precaution in atria.

Figure 16a Open connections and balconies

See section 4.3.2.8

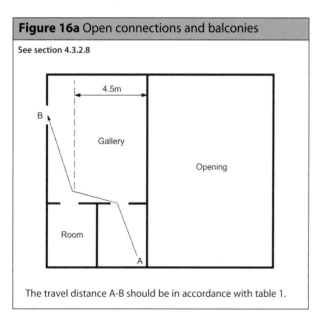

The travel distance A-B should be in accordance with table 1.

Figure 16b Open connections and balconies

See section 4.3.2.8

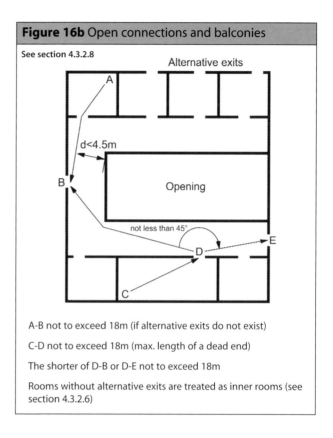

A-B not to exceed 18m (if alternative exits do not exist)

C-D not to exceed 18m (max. length of a dead end)

The shorter of D-B or D-E not to exceed 18m

Rooms without alternative exits are treated as inner rooms (see section 4.3.2.6)

4.3.2.9 Access to storey exits

Any storey which has more than one escape stair should be planned so that it is not necessary to pass through one stairway to reach another. However it would be acceptable to pass through one stairway's protected lobby to reach another stair.

4.3.2.10 Separation of circulation routes from stairways

Unless the doors to a protected stairway and any associated exit passageway are fitted with an automatic release mechanism, the stairway and any associated exit passageway should not form part of the primary circulation route between different parts of the building at the same level. This is because the self-closing fire doors are more likely to be rendered ineffective as a result of their constant use, or because some occupants may regard them as an impediment. For example, the doors are likely to be wedged open or have their closers removed.

4.3.2.11 Height of escape routes

All escape routes should have a clear headroom of not less than 2m except in doorways.

4.3.2.12 Width of escape routes and exits

The width of escape routes and exits depends on the number of persons needing to use them. BB98 (ref) states that: corridors leading to more than one or two teaching spaces should have a clear width (see BS 8300:2001) of at least 1.8m; in new schools this minimum width should be at least 1.9m; if the corridor contains lockers the width should be 2.7m; and, smaller corridors (to not more than one or two teaching spaces) should have a clear width of at least 1200mm – this conforms with Approved Document M *Access to and Use of buildings*.

In dead ends, the minimum corridor width is 1600mm unless a single room of less than 60 people is served by a corridor of less than 4.5m in length, in which case the width should not be less than 1200mm.

For escape purposes, the minimum corridor width of 1200mm, as recommended by AD M, is sufficient if the corridor is not expected to serve as means of escape for more than 250 people. If the number of people is greater than this, the minimum width should be increased by an additional 50mm for each additional 10 persons (or part of 10). However, larger corridor widths should be used where possible, as recommended by BB98.

The aggregate width of all the escape routes should be not less than that required to accommodate the maximum numbers of people likely to use them.

Where the maximum number of people likely to use the escape route and exit is not known, the appropriate capacity should be calculated on the basis of the occupant capacity (see table 2).

Guidance on the spacing of fixed seating for auditoria is given in BS 5588-6:1991.

The distance between a row of seats and the row in front should be at least 305mm. Aisles between blocks of seats should be 1050mm wide, and a block of seats should be no more than 14 seats wide (ie, 7 seats to the nearest aisle).

4.3.2.13 Calculating exit capacity

If a storey or room has two or more storey exits it has to be assumed that a fire might prevent the occupants from using one of them. The remaining exit(s) need to be wide enough to allow all the occupants to leave quickly. Therefore when deciding on the total width of exits needed according to table 4 below, the largest exit should be discounted. This may have implications for the width of stairs, because they should be at least as wide as any storey exit leading onto them. Although some stairs are not subject to discounting (see section 4.4.5.2), the storey exits onto them will be.

Table 4 Escape route width and exit capacity	
Maximum number of persons	Minimum width mm[1][2][3]
60	750[4]
110	850
220	1050
More than 220	50mm for each additional 10 persons or part of 10[5]

Notes:

1. Refer to appendix E on methods of measurement.

2. In order to follow the guidance in BB 98 and AD M, the widths given in the table may need to be increased (see section 4.3.2.12 for corridors, and table 5 for doors)

3. Widths less than 1050mm should not be interpolated.

4. May be reduced to 530mm for gangways between fixed storage racking.

5. 50mm for each additional 10 persons (or part of 10) does not apply to an opening serving less than 220 persons.

AD M contains the following recommendations for the minimum widths of exit doors (see table 5).

Table 5 Minimum effective clear widths of doors, recommended by AD M		
Direction and width of approach	New buildings (mm)	Existing buildings (mm)
Straight on (without a turn or oblique approach)	800	750
At right angles to an access route at least 1500mm wide	800	750
At right angles to an access route at least 1200mm wide	825	775
External doors to buildings used by the general public	1000	775

Note: where there are a large number of pupils with special educational needs (SEN), BB77 and the SSLD for doors (ref) both recommend a clear open width of 900mm.

The total number of persons which two or more available exits (after discounting) can accommodate is found by adding the maximum number of persons that can be accommodated by each exit width. For example, three exits each 850mm wide will accommodate 3 x 110 = 330 persons (not the 510 persons accommodated by a single exit 2550mm wide).

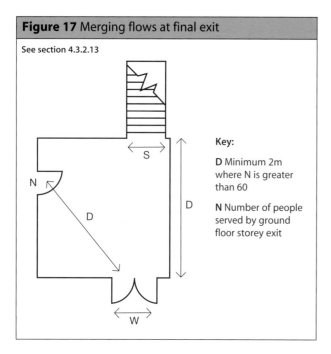

Figure 17 Merging flows at final exit

See section 4.3.2.13

Key:

D Minimum 2m where N is greater than 60

N Number of people served by ground floor storey exit

Where a ground floor storey exit shares a final exit with a stair via a ground floor lobby, the width of the final exit should be sufficient to enable a maximum evacuation flow rate equal to or greater than that from the storey exit and stair combined (see figure 17).

This can be calculated from the following formula:

W = ((N/2.5) + (60S))/80

Where:

W = width of final exit, in metres

N = number of people served by ground floor storey exit

S = stair width in metres

Note: Where the number of persons (N) entering the lobby from the ground floor is more than 60 then the distance from the foot of the stair, or the storey exit, to the final exit should be a minimum of two metres (see figure 17). Where this cannot be achieved then the width of the

final exit (W) should be no less than the width of the stair plus the width of the storey exit.

Worked example

A ground floor storey exit serving 250 persons shares a common final exit with a 1.2 m wide stair.

Required final exit = ((250/2.5) + (1.2 x 60))/80
width (metres) = 2.150 metres.

4.3.2.14 Protected corridors

Dead-end corridors should be avoided. However, where they are present in existing buildings, every dead-end corridor (excluding recesses and extensions not exceeding 2m deep) should be a protected corridor. See also section 4.3.2.1 and 4.3.2.5.

4.3.2.15 Enclosure of corridors that are not protected corridors

Where a corridor that is used as a means of escape, but is not a protected corridor, is enclosed by partitions, those partitions provide some defence against the spread of smoke in the early stages of a fire, even though they may have no fire resistance rating. To maintain this defence the partitions should be carried up to the soffit of the structural floor above, or to a suspended ceiling and openings into rooms from the corridor should be fitted with doors, which need not be fire doors. Open planning, while offering no impediment to smoke spread, has the compensation that occupants can become aware of a fire quickly.

4.3.2.16 Sub-division of corridors

If a corridor provides access to alternative escape routes, there is a risk that smoke will spread along it and make both routes impassable before all occupants have escaped.

To avoid this, every corridor more than 12m long which connects two or more storey exits, should be sub-divided by a self-closing fire door(s) (and any necessary associated screens). The fire door(s) and any associated screen(s) should be positioned approximately mid-way between the two storey exits to effectively safeguard the route from smoke (having regard to the layout of the corridor and to any adjacent fire risks).

Figure 18 Sub-division of corridors

See section 4.3.2.16

a. Section to show use of cavity barriers above the corridor enclosure.

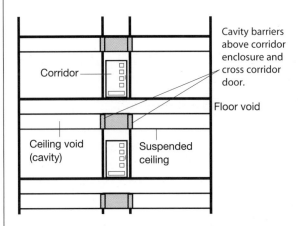

Cavity barriers above corridor enclosure and cross corridor door.

Floor void

Corridor

Ceiling void (cavity)

Suspended ceiling

Where the corridor is a protected escape route, cavity barriers may also be required in any floor void beneath the corridor enclosure. See section 6.4.3.2

b. Plan showing sub-division of storey by fire-resisting construction, as required at a compartment wall.

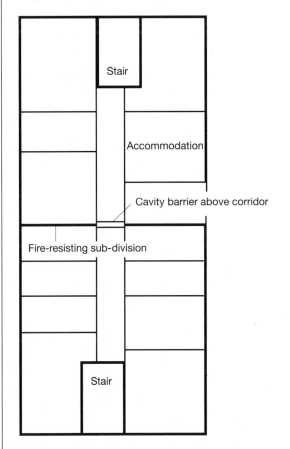

Stair

Accommodation

Cavity barrier above corridor

Fire-resisting sub-division

Stair

The sub-division should be carried full storey height and includes sub-division of the corridor. A cavity barrier may be used in any ceiling void over the sub-division.

Where a cavity exists above the enclosures to any such corridor, because the enclosures are not carried to full storey height or (in the case of a top storey) to the underside of the roof covering, the potential for smoke to bypass the sub-division should be restricted by fitting cavity barriers on the line of the enclosure(s) to and across the corridor.

Any door which could provide a path for smoke to bypass the sub-division should be made self-closing (but need not necessarily be fire-resisting).

If a dead-end portion of a corridor provides access to a point from which alternative escape routes are available, there is a risk that smoke from a fire could make both routes impassable before the occupants in the dead-end have escaped.

To avoid this every dead-end corridor exceeding 4.5m in length should be separated by self-closing fire doors (together with any necessary associated screens) from any part of the corridor which:

a. provides two directions of escape (see figure 19(a)); or

b. continues past one storey exit to another (see figure 19(b)).

Figure 19 Dead-end corridors

See section 4.3.2.16

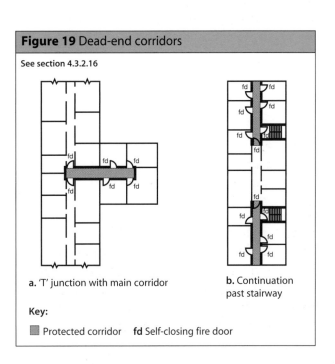

a. 'T' junction with main corridor

b. Continuation past stairway

Key:

 Protected corridor   **fd** Self-closing fire door

245.55

4.3.2.17 External escape routes

Where an external escape route (other than a stair) is beside an external wall of the building, that part of the external wall within 1800mm of the escape route should be of fire-resisting construction, up to a height of 1100mm above the paving level of the route. For guidance on external escape stairs see section 4.4.8.

4.3.2.18 Escape over flat roofs

If more than one escape route is available from a storey, or part of a building, one of those routes may be by way of a flat roof, provided that the route does not serve pupils or the public and:

- the roof should be part of the same building from which escape is being made;

- the route across the roof should lead to a storey exit or external escape route;

- the part of the roof forming the escape route and its supporting structure, together with any opening within 3m of the escape route, should be fire-resisting (see appendix A table A1); and

- the route should be adequately defined and guarded by walls and/or protective barriers which meet the provisions in Approved Document K, *Protection from falling, collision and impact.*

4.4 Design for vertical escape

4.4.1 Introduction

An important aspect of means of escape in multi-storey buildings is the availability of a sufficient number of adequately sized and protected escape stairs. This section deals with escape stairs and includes measures necessary to protect them in all types of building.

The limitation of distances of horizontal travel for means of escape purposes means that most people should be able independently to reach the safety of a protected escape route or final exit. However, some people, for example those who use wheelchairs, may not be able to use stairways without assistance. For them evacuation involving the use of refuges on escape routes and either assistance down (or

up) stairways, or the use of suitable lifts, will be necessary.

This section should be read in conjunction with the general provisions in section 4.5.

4.4.1.1 Exclusions on escape routes

Note that spiral stairs and fixed ladders are precluded from use in schools as part of an escape route for pupils – see section 4.5.4.3 (AD K Stairs Ramps and Guards). Similarly, single steps, apart from in doorways, should be avoided on escape routes. Lifts (except fire-fighting lifts and for a suitably designed and installed evacuation lift that may be used for the evacuation of disabled people in a fire); portable ladders and throw-out ladders; and manipulative apparatus and appliances, eg, fold down ladders and chutes are not deemed to be appropriate.

4.4.2 Number of escape stairs

The number of escape stairs needed in a building (or part of a building) will be determined by:

- the constraints imposed in section 4.3 on the design of horizontal escape routes;

- whether a single stair is acceptable(see section 4.4.2.1); and

- provision of adequate width for escape (see section 4.4.4) while allowing for the possibility that a stair may have to be discounted because of fire or smoke (see section 4.4.5.2).

In larger buildings, provisions for access for the Fire and Rescue Service may apply, in which case, some escape stairs may also need to serve as fire-fighting stairs. The number of escape stairs may therefore be affected by provisions made in section 8.4.3.

4.4.2.1 Single escape stairs

Ideally, single stairways should be avoided in new buildings. This is consistent with the avoidance of dead ends in new buildings.

The situations where a building (or part of a building) may be served by a single escape stair are:

a. from a basement which is allowed to have a single escape route in accordance with section 4.3.2.1 and table 1; and

b. from a building which has no storey with a floor level more than 11m above ground level and in which every storey is allowed to have a single escape route in accordance with section 4.3.2.1 and table 1.

Note: where case (b) applies, the storeys above the first floor level should only be occupied by adults. There should be no more than 120 pupils plus supervisors on the first storey and no place of special fire hazard. Classrooms and stores should not open onto the stairway.

4.4.3 Provision of refuges

During the course of a fire it is now recommended that mobility-impaired people are directed to a place of safety/refuge until they can be escorted out of the building; there shall be a place of refuge on each protected stairway. Note that the expectation is that all disabled occupants will have left the building before the fire fighters arrive.

Refuges are relatively safe waiting areas for short periods. They are not areas where disabled people should be left alone indefinitely until rescued by the Fire and Rescue Service, or until the fire is extinguished.

A refuge should be provided for each protected stairway affording egress from each storey,

except storeys consisting exclusively of plant rooms. All glazing within refuges should have appropriate levels of integrity and insulation fire resistance.

Note: Whilst a refuge should be provided for each stairway, they need not necessarily be located within the stair enclosure but should enable direct access to the stair. The number of refuge spaces need not necessarily equal the sum of the number of wheelchair users who can be present in the building. Refuges form a part of the management plan and it may be that more than one disabled person will use a single refuge as they pass through as a part of the evacuation procedure.

The following are examples of satisfactory refuges:

- an enclosure such as a compartment (see figure 20), protected lobby, protected corridor or protected stairway (see figure 21); and

- an area in the open air such as a flat roof, balcony, podium or similar place which is sufficiently protected (or remote) from any fire risk and provided with its own means of escape.

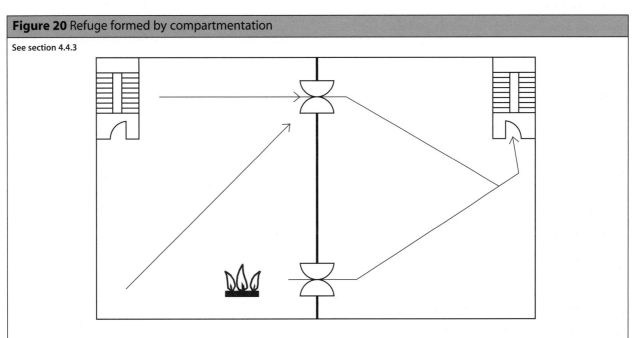

Figure 20 Refuge formed by compartmentation

See section 4.4.3

Storey divided into two refuges by compartment wall (stairways not provided with wheelchair space).

Note: Persons occupying the left hand compartment would not reach a refuge until they had entered the right-hand compartment. Two doorsets in the partition are necessary in case access to one of the doorsets is blocked by fire.

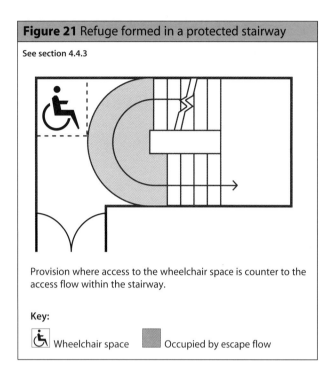

Figure 21 Refuge formed in a protected stairway

See section 4.4.3

Provision where access to the wheelchair space is counter to the access flow within the stairway.

Key:

Wheelchair space   ▨ Occupied by escape flow

Each refuge should provide an area accessible to a wheelchair of at least 900mm x 1400mm in which a wheelchair user can await assistance. Where a refuge is a protected stairway or protected lobby or protected corridor, the wheelchair space should not reduce the width of the escape route. Where the wheelchair space is within a protected stairway, access to the wheelchair space should not obstruct the flow of persons escaping.

Refuges and evacuation lifts should be clearly identified by appropriate fire safety signs. Where a refuge is in a lobby or stairway the sign should be accompanied by a blue mandatory sign worded 'Refuge – keep clear'.

4.4.3.1   Communication

To facilitate the effective evacuation of people from refuges an emergency voice communication (EVC) system should be provided. It is essential that the occupants of each refuge are able to alert other people that they are in need of assistance and for them to be reassured that this assistance will be forthcoming.

The EVC system should comply with BS 5839-9:2003 and consist of Type B outstations which communicate with a master station located in the building control room (where one exists) or adjacent to the fire alarm panel.

In some buildings it may be more appropriate to use an alternative approach such as the use of wireless technology.

4.4.4   Width of escape stairs

The width of escape stairs should:

- be at least 1100mm;
- be not less than the width(s) required for any exit(s) affording access to them;
- conform with the minimum widths given in table 6; and
- not reduce in width at any point on the way to a final exit.

In order to follow the guidance in AD M, BB77 and the SSLD for stairs (ref), the minimum widths may need to be increased. For example, the SSLD for stairs sets a minimum width of 1600mm. Note however that the SSLD only applies to new secondary schools.

If the resultant width of the stair is more than 1800mm, then for reasons of safety in normal use, the guidance in Approved Document K *Protection from falling, collision and impact,* is that the stair should have a central handrail. In such a case the stair width on each side of the central handrail needs to be considered separately for the purpose of assessing stair capacity. Coupled with the requirement that no stair should be less than 1100mm wide, this means there will be no stair between 1800mm and 2200mm wide.

A second handrail at low height should be provided for younger children.

Where an exit route from a stair also forms the escape from the ground and/or basement storeys, the width may need to be increased accordingly (see section 4.3.2.13).

4.4.5   Calculation of minimum stair width

The capacity of stairs and exits will need to be calculated based upon the premise that at least one of the alternatives will become smoke-logged and so everyone will have to use the remaining exits.

4.4.5.1   General

Every escape stair should be wide enough to accommodate the number of persons needing

4.4.5.3 Stair Capacity

The escape stairs (in conjunction with the rest of the means of escape) should have the capacity to allow all floors to be evacuated simultaneously. In calculating the width of the stairs account is taken of the number of people temporarily housed in the stairways during the evacuation.

The minimum width of a stair is 1100mm, serving up to 220 people. When designing for greater numbers of people, the capacity of stairs of widths from 1100 to 1800mm is given in table 6. Note that these stair widths assume a simultaneous evacuation of the entire building. Phased evacuation is not considered appropriate for school buildings.

As an alternative to using table 6, the capacity of stairs 1100mm or wider (for simultaneous evacuation) can be derived from the formula:

$$p = 200w + 50(w - 0.3)(n - 1)$$

or

$$w = \frac{P + 15n - 15}{150 + 50n}$$

where:

(P) is the number of people that can be served;
(w) is the width of the stair, in metres; and
(n) is the number of storeys served.

Note 1: Separate calculations should be made for stairs/flights serving basement storeys and those serving upper storeys.

Note 2: The population 'P' should be divided by the number of available stairs.

to use it in an emergency. This width will depend on the number of stairs provided.

As with the design of horizontal escape routes, where the maximum number of people needing to use the escape stairs is not known, the occupant capacity should be calculated on the basis of the appropriate floor space factors. Guidance for this is set out in table 2.

4.4.5.2 Discounting of stairs

Where two or more stairs are provided it should be assumed that one of them might not be available due to fire. It is therefore necessary to discount each stair in turn in order to ensure that the capacity of the remaining stair(s) is adequate for the number of persons needing to escape. The stair discounting rule applies to a building fitted with a sprinkler system.

Two exceptions to the above discounting rules are if the escape stairs:

- are protected by a smoke control system designed in accordance with BS EN 12101-6:2005; and
- are approached on each storey through a protected lobby (a protected lobby need not be provided on the topmost storey for the exception still to apply).

In such cases the likelihood of a stair not being available is significantly reduced and it is not necessary to discount a stair. However, a storey exit still needs to be discounted, see section 4.3.2.13.

Table 6 Capacity of a stair for basements and for simultaneous evacuation of the building								
No of floors served	Maximum number of persons served by a stair of width							
	1100mm	1200mm	1300mm	1400mm	1500mm	1600mm	1700mm	1800mm
1	220	240	260	280	300	320	340	360
2	260	285	310	335	360	385	410	435
3	300	330	360	390	420	450	480	510
4	340	375	410	445	480	515	550	585
5	380	420	460	500	540	580	620	600
6	420	465	510	555	600	645	690	735

* Should it be necessary to go above six floors the designer is recommended to use guidance in AD B.

Note 3: The formula is particularly useful when determining the width of stairs serving a building (or part of a building) where the occupants are not distributed evenly – either within a storey or between storeys.

Note 4: In the formula, the first part (200w) represents the number of persons estimated to have left the stair after 2.5 minutes of evacuation. The second part (50 (w-0.3) (n-1)) represents the number of persons estimated to be accommodated on the stair after this time.

Example

Consider a school with 4 storeys, 4 stairs, containing 1680 pupils and staff. The number of floors served is 3 (the fourth storey is the ground floor). Assuming an even distribution of pupils and staff over the four storeys, there will be 420 people per floor.

If it is assumed that one stair is unavailable due to smoke-logging, the 1260 people not on the ground floor will have to use the three remaining stairs, ie, 420 people per stair. Substituting P = 420 and n = 3 into the formula for w gives

$$w = \frac{420 + 15 \times 3 - 15}{150 + 50 \times 3} = \frac{450}{300} = 1.5$$

If all the stairs have smoke control or lobby protection, it is reasonable to assume that they will not become smoke-logged in the event of a fire. However, a storey exit should still be discounted since the location of the fire on a storey may make an exit inaccessible. Assuming the fire is not on the ground floor, the number of people using the stair will be:

- 420/3 = 140 people per stair from the fire floor; and
- 420/4 = 105 people per stair from each of the other two floors.

ie, a total of 350 people per stair.

Using the formula for w, this gives:

$$w = \frac{350 + 15 \times 3 - 15}{150 + 50 \times 3} = \frac{380}{300} = 1.267$$

The use of smoke control or lobby protection means that narrower stairs can be used.

4.4.6   Protection of escape stairs

4.4.6.1   General

Escape stairs need to have a satisfactory standard of fire protection if they are to fulfil their role as areas of relative safety during a fire evacuation. The guidance in section 4.4.6.2 should be followed to achieve this.

One of the benefits of protecting stairways from the effects of fire is to allow travel distance to be measured from the furthest point on the relevant floor to the nearest storey exit rather than the final exit of the building.

4.4.6.2   Enclosure of escape stairs

Every internal escape stair should be a protected stairway (ie, it should be within a fire-resisting enclosure).

However an unprotected stair (eg, an accommodation stair) may form part of an internal route to a storey exit or final exit, provided that the distance of travel and the number of people involved are very limited.

4.4.6.3   Access lobbies and corridors

There are situations where an escape stair needs the added protection of a protected lobby or protected corridor. These are:

- where the stair is the only one serving a building (or part of a building) which has more than one storey above or below the ground storey;
- where the stair is a fire-fighting stair; or
- where the option in section 4.4.5.2 (second bullet of second paragraph) has been used so as not to discount one stairway when calculating stair widths.

In these cases protected lobbies or protected corridors are needed at all levels, except the top storey and at all basement levels. Alternatively a smoke control system may be used, unless the stair is a fire-fighting stair, to ensure the stairway remains smoke free.

A protected lobby should also be provided between an escape stairway and a place of special fire hazard. In this case, the lobby should have not less than 0.4m^2 of permanent ventilation.

4.4.6.4 Exits from protected stairways

Every protected stairway should discharge directly to a final exit; or by way of a protected exit passageway to a final exit. Any such protected exit passageway should have the same standard of fire resistance and lobby protection as the stairway it serves.

The exit from a protected stairway should meet the provisions in section 4.5.5.4.

4.4.6.5 Separation of adjoining stairways

Where two protected stairways are adjacent, they and any protected exit passageways linking them to final exits, should be separated by an imperforate barrier.

4.4.6.6 Use of space within protected stairways

A protected stairway needs to be free of potential sources of fire. Consequently, facilities that may be incorporated in a protected stairway are limited to the following:

a. sanitary accommodation or washrooms, so long as the accommodation is not used as a cloakroom. A gas water heater or sanitary towel incinerator may be installed in the accommodation but not any other gas appliance;

b. a lift well may be included in a protected stairway, if it is not a fire-fighting stair (see 8.4.1);

c. a reception desk or enquiry office area at ground or access level, if it is not in the only stair serving the building or part of the building. The reception or enquiry office area should not be more than 10m² in area; and/or

d. cupboards enclosed with fire-resisting construction, if they are not in the only stair serving the building or part of the building.

4.4.6.7 External walls of protected stairways

With some configurations of external wall, a fire in one part of a building could subject the external wall of a protected stairway to heat (for example, where the two are adjacent at an internal angle in the facade as shown in figure 22). If the external wall of the protected stairway has little fire resistance, there is a risk that this could prevent the safe use of the stair.

Therefore, if:

a. a protected stairway projects beyond, or is recessed from, or is in an internal angle of, the adjoining external wall of the building; then

b. the distance between any unprotected area in the external enclosures to the building and any unprotected area in the enclosure to the stairway should be at least 1800mm (see figure 22).

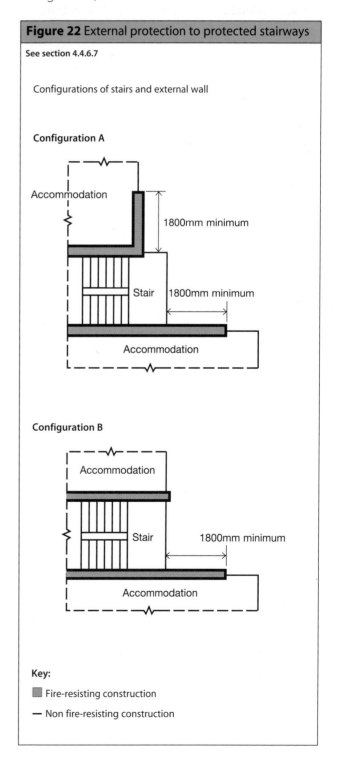

Figure 22 External protection to protected stairways

See section 4.4.6.7

Configurations of stairs and external wall

Configuration A

Accommodation

1800mm minimum

Stair   1800mm minimum

Accommodation

Configuration B

Accommodation

Stair   1800mm minimum

Accommodation

Key:

▨ Fire-resisting construction

— Non fire-resisting construction

4.4.7 Basement stairs

Because of their situation, basement stairways are more likely to be filled with smoke and heat than stairs in ground and upper storeys.

Special measures are therefore needed in order to prevent a basement fire endangering upper storeys.

If an escape stair forms part of the only escape route from an upper storey of a building (or part of a building) it should not be continued down to serve any basement storey. The basement should be served by a separate stair.

If there is more than one escape stair from an upper storey of a building (or part of a building), only one of the stairs serving the upper storeys of the building (or part) need be terminated at ground level. Other stairs may connect with the basement storey(s) if there is a protected lobby, or a protected corridor between the stair(s) and accommodation at each basement level.

Figure 23 Fire resistance to areas adjacent to external stairs

See section 4.4.8

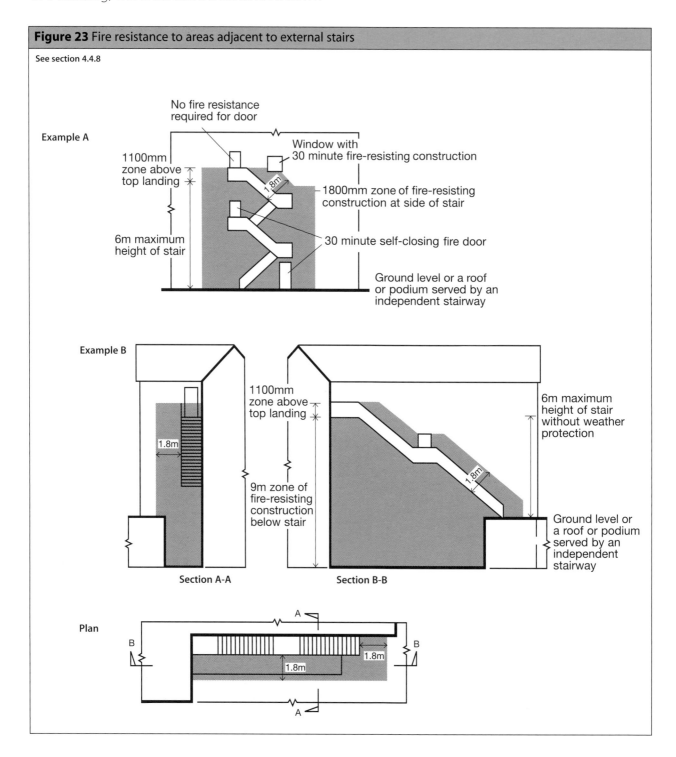

4.4.8   External escape stairs

If more than one escape route is available from a storey (or part of a building), some of the escape routes from that storey or part of the building may be by way of an external escape stair, provided that:

- there is at least one internal escape stair from every part of each storey (excluding plant areas); and
- the route is not intended for use by pupils or members of the public; or
- the route does not extend above the first floor.

Where external stairs are acceptable as forming part of an escape route, they should meet the following provisions:

- all doors giving access to the stair should be fire-resisting and self-closing, except that a fire-resisting door is not required at the head of any stair leading downwards where there is only one exit from the building onto the top landing;
- any part of the external envelope of the building within 1800mm of (and 9m vertically below), the flights and landings of an external escape stair should be of fire-resisting construction, except that the 1800mm dimension may be reduced to 1100mm above the top level of the stair if it is not a stair up from a basement to ground level (see figure 23);
- there is protection by fire-resisting construction for any part of the building (including any doors) within 1800mm of the escape route from the stair to a place of safety, unless there is a choice of routes from the foot of the stair that would enable the people escaping to avoid exposure to the effects of the fire in the adjoining building;
- any stair more than 6m in vertical extent is protected from the effects of adverse weather conditions. (This should not be taken to imply a full enclosure. Much will depend on the location of the stair and the degree of protection given to the stair by the building itself); and
- glazing in areas of fire-resisting construction mentioned above should also be fire-resisting and normally integrity performance or insulation performance where a risk assessment indicates that the potential fire hazard requires protection against heat in the post flashover phase.

4.5   General provisions for means of escape

4.5.1   Introduction

This section gives guidance on the construction and protection of escape routes generally, service installations and other matters associated with the design of escape routes.

It should therefore be read in conjunction with sections 4.3 and 4.4.

4.5.2   Protection of escape routes

If a fire starts, then however it spreads within the building it must not compromise the use of dedicated, protected escape routes. Such routes are protected by elements of construction that are able to satisfy the appropriate criteria of the fire resistance test, see appendix A.

4.5.2.1   Fire resistance of enclosures

Other elements of the structure may be required to provide fire resistance for the purpose of:

- reducing the area of the building that is at risk, in line with regulatory recommendations;
- protecting against disproportionate damage to the facility in the event of a fire;
- containing areas of high risk; and
- protecting areas that are critical to the running of the school, or have a high intrinsic value.

Details of fire resistance test criteria and standards of performance, are set out in appendix A. Generally, a 30-minute standard is sufficient for the protection of means of escape. The exceptions to this are when greater fire resistance is required by the guidance on Requirements B3 or B5, or some other specific instance to meet Requirement B1 (see section 4.3).

All walls, partitions and other enclosures that need to be fire-resisting to meet the provisions in this document (including roofs that form part of a means of escape), should have the appropriate performance given in tables A1 and A2 of appendix A.

Glazed elements protecting a means of escape should meet limitations defined for the use of integrity glass (see section 4.5.2.3).

4.5.2.2   Fire resistance of doors

Details of fire resistance test criteria and standards of performance, are set out in appendix C.

All doors that need to be fire-resisting to meet the provisions in this document should have the appropriate performance given in table C1 of appendix C.

Doors should also meet limitations defined for integrity glass (see section 4.5.2.3).

4.5.2.3 Fire resistance of glazed elements

Where glazed elements in fire-resisting enclosures and doors are only able to satisfy the relevant performance in terms of integrity, the use of glass is limited. These limitations depend on whether the enclosure forms part of a protected shaft (see section 6.3) and the provisions set out in appendix A, table A4.

Where the relevant performance can be met in terms of both integrity and insulation, there is no restriction in this document on the use or amount of glass.

Attention is also drawn to the guidance on the safety of glazing in Approved Document N, *Glazing – safety in relation to impact, opening and cleaning*.

4.5.3 Doors on escape routes

The time taken to negotiate a closed door can be critical in escaping. Doors on escape routes (both within and from the building) should therefore be readily openable, if undue delay is to be avoided. Accordingly the provisions in section 4.5.3.1 – 4.5.3.5 should be met.

4.5.3.1 Door fastenings

In general, doors on escape routes (whether or not the doors are fire doors), should either not be fitted with lock, latch or bolt fastenings, or they should only be fitted with simple fastenings that can be readily operated from the side approached by people making an escape. The operation of these fastenings should be readily apparent; without the use of a key and without having to manipulate more than one mechanism. This is not intended to prevent doors being fitted with hardware to allow them to be locked when the rooms are empty.

Where a door on an escape route has to be secured against entry when the building or part of the building is occupied, it should only be fitted with a lock or fastening which is readily operated, without a key, from the side approached by people making their escape. Similarly, where a secure door is operated by a code, combination, swipe or proximity card,

biometric data or similar means, it should also be capable of being overridden from the side approached by people making their escape.

Electrically powered locks should return to the unlocked position:

- on operation of the fire alarm system;
- on loss of power or system error; and
- on activation of a manual door release unit (Type A) to BS EN 54-11:2001 positioned at the door on the side approached by people making their escape. Where the door provides escape in either direction, a unit should be installed on both sides of the door.

Doors on escape routes from rooms with an occupant capacity of more than 60 should either not be fitted with lock, latch or bolt fastenings, or be fitted with panic fastenings in accordance with BS EN 1125:1997.

It may also be appropriate to accept on some final exit doors locks for security that are used only when the building is empty. In these cases the emphasis for the safe use of these locks must be placed on management procedures.

Guidance about door closing and 'hold-open' devices for fire doors is given in appendix C.

4.5.3.2 Direction of opening

The door of any doorway or exit should, if reasonably practicable, be hung to open in the direction of escape and should always do so if the number of persons that might be expected to use the door at the time of a fire is more than 60.

Note: Where there is a very high fire risk with potential for rapid fire growth, doors should open in the direction of escape even where the number of persons does not exceed 60.

4.5.3.3 Amount of opening and effect on associated escape routes

All doors on escape routes should be hung to open not less than 90 degrees with a swing that is clear of any change of floor level, other than a threshold or single step on the line of the doorway (see section 4.5.4.2) and which does not reduce the effective width of any escape route across a landing.

A door that opens towards a corridor or a stairway should be sufficiently recessed to prevent its swing from encroaching on the effective width of the stairway or corridor.

4.5.3.4 Vision panels in doors

Vision panels are needed where doors on escape routes sub-divide corridors, or where any doors are hung to swing both ways. Note also the provision in Approved Document M, *Access to and Use of buildings*, concerning vision panels in doors across accessible corridors and passageways and the provisions for the safety of glazing in Approved Document N, *Glazing – safety in relation to impact, opening and cleaning*.

Vision panels in insulating doors should be of insulating fire-resisting glass, in which case the area of use of the insulating glass does not need to be limited. If uninsulating fire-resistant glazing is used, the vision panels should be less than 10% of the door area, and not be less than 500mm above the floor.

4.5.3.5 Revolving and automatic doors

Revolving doors, automatic doors and turnstiles can obstruct the passage of persons escaping. Accordingly, they should not be placed across escape routes unless:

- they are to the required width and are automatic doors and either they:
  i. are arranged to fail safely to outward opening from any position of opening; or
  ii. are provided with a monitored failsafe system for opening the doors if the mains supply fails; or
  iii. they fail safely to the open position in the event of power failure;

- non-automatic swing doors of the required width are provided immediately adjacent to the revolving or automatic door or turnstile.

4.5.4 Stairs

4.5.4.1 Construction of escape stairs

The flights and landings of every escape stair should be constructed of materials of limited combustibility in the following situations:

a. if it is the only stair serving the building, or part of the building;

b. if it is within a basement storey;

c. if it is external (see section 4.4.8); or

d. if it is a fire-fighting stair (see section 8.4).

Note: In satisfying the above conditions, combustible materials may be added to the horizontal surface of these stairs (except in the case of fire-fighting stairs).

There is further guidance on the construction of fire-fighting stairs in section 8.4. Dimensional constraints on the design of stairs generally, to meet requirements for safety in use, are given in Approved Document K, *Protection from falling, collision and impact*.

The preferred rise for each step should be not more than 150mm, and the tread between 250mm and 280mm (not less than the lower value), for steps in general, and that there should be between 3 and 16 treads per flight. The length of any landing on a staircase should be at least the width of the stair, and there should be a change of direction at least every two flights.

4.5.4.2 Single steps

Single steps may cause falls and should only be used on escape routes where they are prominently marked. A single step on the line of a doorway is acceptable, subject to section 4.5.5.4.

The rise should be not more than 150mm, and the tread not less than 300mm, for steps providing for small changes in level. There should be a level stretch of corridor extending for 1.5m either side of any such steps.

4.5.4.3 Helical stairs, spiral stairs and fixed ladders

Helical stairs, spiral stairs and fixed ladders may form part of an escape route, provided they are not intended to serve pupils or members of the public, subject to the following restrictions:

a. helical and spiral stairs should be designed in accordance with BS 5395-2:1984; and

b. fixed ladders should only be intended for use in circumstances where it is not practical to provide a conventional stair, for example, as access to plant rooms that are not normally occupied.

Guidance on the design of helical and spiral stairs and fixed ladders, from the aspect of safety in use, is given in Approved Document K, *Protection from falling, collision and impact*.

4.5.5 General

4.5.5.1 Headroom in escape routes

All escape routes should have a clear headroom of not less than 2m and there should be no projection below this height (except for door frames).

4.5.5.2 Floors of escape routes

The floorings of all escape routes (including the treads of steps and surfaces of ramps and landings) should be chosen to minimise their slipperiness when wet.

4.5.5.3 Ramps and sloping floors

Where a ramp forms part of an escape route it should meet the provisions in Approved Document M, *Access to and Use of buildings*. Any sloping floor or tier should be constructed with a pitch of not more than 35° to the horizontal.

Inclines and ramps may be preferable to stairs, particularly where the change in level is slight, or where wheelchair access is a requirement. The gradient of the incline or ramp should not exceed 1:12. AD M has further details.

Further guidance on the design of ramps and associated landings and on aisles and gangways in places where there is fixed seating, from the aspect of safety in use, is given in AD K and AD M. The design of means of escape in places with fixed seating is dealt with in section 4.3.2.12 by reference to BS 5588-6:1991.

4.5.5.4 Final exits

Final exits need to be dimensioned and sited to facilitate the evacuation of persons out of and away from the building. Accordingly, they should be not less in width than the minimum width required for the escape route(s) they serve and should also meet the conditions below.

Final exits should be sited to ensure rapid dispersal of persons from the vicinity of the building so that they are no longer in danger from fire and smoke. Direct access to a street, passageway, walkway or open space should be available. The route clear of the building should be well defined and, if necessary, have suitable guarding.

The designers shall ensure adequate provision is made for the safe assembly of the school occupants, for example:

- Final exits to an enclosed courtyard are not suitable.
- Assembly points should not get in the way of access for the Fire and Rescue Service.
- Assembly points should not be at risk from breaking glass falling from windows of the building façade.

Escape routes from facilities such as indoor swimming pools and games areas will need to allow for possible egress in cold weather as well as day and night-time conditions.

Final exits should not present an obstacle to wheelchair users and other people with disabilities. Where a final exit is accessed without the need to first traverse steps then a level threshold and, where necessary, a ramp should be provided.

Final exits need to be apparent to persons who may need to use them. This is particularly important where the exit opens off a stair that continues down, or up, beyond the level of the final exit.

Final exits should be sited so that they are clear of any risk from fire or smoke in a basement (such as the outlets to basement smoke vents, see section 8.5), or from openings to transformer chambers, refuse chambers, boiler rooms and similar risks.

4.5.5.5 Lighting of escape routes

All escape routes should have adequate artificial lighting. Routes and areas apart from the exceptions listed below should also have escape lighting which illuminates the route if the main supply fails. The exceptions are:

a. accommodation open on one side to view sport or entertainment during normal daylight hours;

b. parts of school buildings with natural light and used only during normal school hours; and

c. toilet accommodation with a window and with a floor area not more than 8m².

Lighting to escape stairs should be on a separate circuit from that supplying any other part of the escape route.

Standards for the installation of a system of escape lighting are given in BS 5266-1:2005.

4.5.5.6 Exit signs

Every escape route (other than those in ordinary use) should be distinctively and conspicuously marked by emergency exit sign(s) of adequate size complying with the *Health and Safety (Safety signs and signals) Regulations, 1996*. In general, signs containing symbols or pictograms which conform to BS 5499-1:2002, satisfy these regulations. In some buildings additional signs may be needed to meet requirements under other legislation.

Suitable signs should also be provided for refuges (see section 4.4.3).

Note: Advice on fire safety signs, including emergency escape signs, is given in an HSE publication: *Safety Signs and Signals: Guidance on Regulations*. Further guidance is given in RRO Guide Part 2, section 6 on safety signs and notices.

4.5.5.7   Protected power circuits

Where it is critical for electrical circuits to be able to continue to function during a fire, protected circuits are needed. The potential for damage to cables forming protected circuits should be limited by the use of sufficiently robust cables, careful selection of cable routes and/or by the provision of physical protection in areas where cables may be susceptible to damage. Methods of cable support should generally be non-combustible and such that circuit integrity will not be reduced below that afforded by the cable.

A protected circuit for operation of equipment in the event of fire should consist of cable meeting at least the requirements for PH 30 classification when tested in accordance with BS EN 50200:2006 (incorporating appendix E of that standard), or an equivalent standard. It should follow a route selected to pass only through parts of the building in which the fire risk is negligible and should be separate from any circuit provided for another purpose.

In large or complex buildings there may be fire protection systems, eg, sprinkler systems, that need to operate for an extended period during a fire. Further guidance on the selection of cables for such systems is given in BS 5839-1, BS 5266-1 and BS 7346-6.

4.5.6   Lifts

4.5.6.1   Evacuation lifts

In general it is not appropriate to use lifts when there is a fire in the building because there is always the danger of people being trapped in a lift that has become immobilised as a result of the fire. However, in some circumstances a lift may be provided as part of a management plan for evacuating people. In such cases the lift installation may need to be appropriately sited and protected and may need to contain a number of safety features that are intended to ensure that the lift remains usable for evacuation purposes during the fire. Guidance on the design and use of evacuation lifts is given in BS 5588-8:1999.

As schools have become more inclusive, the need for lifts to assist those pupils with limited mobility is increasingly widespread. It may be beneficial to design all lifts to be used as evacuation lifts, which will assist the safe escape of anyone with a mobility problem. AD M gives details of minimum lift dimensions, ie, 1100mm wide and 1400mm deep.

4.5.6.2   Fire protection of lift installations

Because lifts connect floors, there is the possibility that they may prejudice escape routes. To safeguard against this, the conditions below should be met.

Lifts, such as wall-climber or feature lifts which rise within a large volume, such as a mall or atrium, and do not have a conventional well, may be at risk if they run through a smoke reservoir. In which case, care is needed to maintain the integrity of the smoke reservoir and protect the occupants of the lift.

Lift wells should be either:

a. contained within the enclosures of a protected stairway; or

b. enclosed throughout their height with fire-resisting construction if they are sited so as to prejudice the means of escape.

A lift well connecting different compartments should form a protected shaft (see section 6.3).

In basements the lift should be approached only by a protected lobby (or protected corridor), unless it is within the enclosure of a protected stairway.

A lift shaft should not be continued down to serve any basement storey if it is:

a. in a building (or part of a building) served by only one escape stair and smoke from a basement fire would be able to prejudice the escape routes in the upper storeys; or

b. within the enclosure to an escape stair which is terminated at ground level.

If the lift well is within a protected stairway which is the only stairway serving the building (or part of the building), then lift motors should not be located in the stairway (to avoid smoke spread from a fire in the lift machinery). However, lift machine rooms, if required, can be sited over the lift well.

4.5.7 Mechanical ventilation and air-conditioning systems

Any system of mechanical ventilation should be designed to ensure that, in a fire, the ductwork does not assist in transferring fire and smoke through the building and put at risk the protected means of escape from the accommodation areas. Any exhaust points should be sited so as not to further jeopardize the building, ie, away from final exits, combustible building cladding or roofing materials and openings into the building.

Ventilation ducts supplying or extracting air directly to or from a protected escape route, should not also serve other areas. A separate ventilation system should be provided for each protected stairway. Where the ductwork system serves more than one part of a sub-divided (see section 4.3.2.16) escape route, a fire damper should be provided where ductwork enters each section of the escape route operated by a smoke detector or suitable fire detection system (see also section 6.5). The fire dampers should close when smoke is detected.

Ducts passing through the enclosure of a protected escape route should be fire-resisting, ie, the ductwork should be constructed in accordance with Method 2 or Method 3 (see section 6.5).

Note: Fire dampers activated only by fusible links are not suitable for protecting escape routes. However an ES classified fire and smoke damper which is activated by a suitable fire detection system may be used (see section 6.5.3.1).

In the case of a system which recirculates air, smoke detectors should be fitted in the extract ductwork before the point of separation of the recirculated air and the air to be discharged to the open air and before any filters or other air cleaning equipment. Such detector(s) should:

a. cause the system to immediately shut down; or

b. switch the ventilation system from recirculating mode to extraction to open air, so as to divert any smoke to the outside of the building.

Mechanical ventilation, unless specifically designed for smoke extraction, should be shut down as soon as smoke is detected in the duct, or upon operation of the fire alarm.

Non-domestic kitchens and plant rooms should have separate and independent extraction systems and the extracted air should not be recirculated.

Guidance on the use of mechanical ventilation is given in BS 5588-6:1991.

Where a pressure differential system is installed, ventilation and air-conditioning systems in the building should be compatible with it when operating under fire conditions.

Further guidance on the design and installation of mechanical ventilation and air conditioning plant is given in BS 5720:1979. Guidance on the provision of smoke detectors in ventilation ductwork is given in BS 5839-1:2002.

Note: sections 6.3.5.5 and 6.5.3 also deal with ventilation and air-conditioning ducts.

4.5.8 Refuse and recycling storage

In order to reduce the likelihood of arson fires, external refuse and recycling stores should be well away from main buildings, and access to them should be restricted. For the purposes of property protection, combustible materials should not be stored within 10m of the outside of the building.

Refuse and recycling storage chambers, refuse and recycling chutes and refuse and recycling hoppers should be sited and constructed in accordance with BS 5906 *Code of practice for storage and on-site treatment of solid waste from buildings*.

Refuse and recycling rooms provided for the storage of refuse recycling should be separated from other parts of the building by fire-resisting construction; and not be located within protected stairways or protected lobbies.

Access to refuse and recycling storage chambers should not be sited adjacent to escape routes or final exits.

Section 5
Internal fire spread (linings)

5.1   Overview

5.1.1   Requirement B2 of the Building Regulations

(1) To inhibit the spread of fire within the building, the internal linings shall:

(a) adequately resist the spread of flame over their surfaces; and

(b) have, if ignited, a rate of heat release or a rate of fire growth, which is reasonable in the circumstances.

(2) In this section 'internal linings' mean the materials or products used in lining any partition, wall, ceiling or other internal structure.

5.1.2   Performance

The Requirements of B2 will be met if the spread of flame over the internal linings of the building is restricted by making provision for them to have low rates of surface spread of flame and, in some cases, to have a low rate of heat release, so as to limit the contribution that the fabric of the building makes to fire growth. In relation to the European fire tests and classification system, the requirements of B2 will be met if the heat released from the internal linings is restricted by making provision for them to have a resistance to ignition and a rate of fire growth which are reasonable in the circumstances.

The extent to which this is necessary is dependent on the location of the lining. Lining materials and finishes should be robust in order to retain good reaction to fire properties.

5.1.3   Introduction

5.1.3.1   Fire spread and lining materials

The choice of materials for walls and ceilings can significantly affect the spread of a fire and its rate of growth, even though they are not likely to be the materials first ignited.

It is particularly important in circulation spaces where linings may offer the main means by which fire spreads and where rapid spread is most likely to prevent occupants from escaping.

Several properties of lining materials influence fire spread. These include the ease of ignition and the rate at which the lining material gives off heat when burning. The guidance relating to the European fire tests and classification provides for control of internal fire spread through control of these properties. This document does not give detailed guidance on other properties such as the generation of smoke and fumes.

5.1.3.2   Floors and stairs

The provisions do not apply to the upper surfaces of floors and stairs because they are not significantly involved in a fire until well developed and thus do not play an important part in fire spread in the early stages of a fire that are most relevant to the safety of occupants.

However, it should be noted that the construction of some stairs and landings is controlled under section 4.5.4.1.

5.1.3.3   Other controls on internal surface properties

There is also guidance on the control of flame spread inside buildings in two other sections. In section 6.4 there is guidance on surfaces exposed in concealed spaces above fire-protecting suspended ceilings and in section 6.5 on enclosures to above ground drainage system pipes.

Note: External flame spread is dealt with in sections 7.2 to 7.4; the fire behaviour of insulating core panels used for internal structures is dealt with in appendix B.

5.1.3.4   Furniture and fittings

Furniture and fittings can have a major effect on fire spread but it is not possible to control them through Building Regulations and they are not dealt with in this document. Fire characteristics of furniture and fittings may be controlled in some buildings under legislation that applies to a building in use, such as licensing conditions.

5.1.3.5 Classification of performance

Appendix A describes the different classes of performance and the appropriate methods of test.

The National classifications used are based on tests in BS 476 *Fire tests on building materials and structures*, namely Part 6: *Method of test for fire propagation for products* and Part 7: *Method of test to determine the classification of the surface spread of flame of products*. However, Part 4: *Non-combustibility test for materials* and Part 11: *Method for assessing the heat emission from building products* are also used as one method of meeting Class 0. Other tests are available for classification of thermoplastic materials if they do not have the appropriate rating under BS 476-7 and three ratings, referred to as TP(a) rigid and TP(a) flexible and TP(b), are used.

The European classifications are described in BS EN 13501-1:2002, *Fire classification of construction products and building elements*, Part 1 – *Classification using data from reaction to fire tests*. They are based on a combination of four European test methods, namely:

BS EN ISO 1182:2002 *Reaction to fire tests for building products – Non combustibility test*;

BS EN ISO 1716:2002 *Reaction to fire tests for building products – Determination of the gross calorific value*;

BS EN 13823:2002 *Reaction to fire tests for building products – Building products excluding floorings exposed to the thermal attack by a single burning item*; and

BS EN ISO 11925-2:2002, *Reaction to fire tests for building products. Ignitability when subjected to direct impingement of flame*.

For some building products, there is currently no generally accepted guidance on the appropriate procedure for testing and classification in accordance with the harmonised European fire tests. Until such a time that the appropriate European test and classification methods for these building products are published, classification may only be possible using existing national test methods.

Table A8, in appendix A, gives typical performance ratings which may be achieved by some generic materials and products.

5.2 Wall and ceiling linings

5.2.1.1 Wall and ceiling linings

Subject to the variations and specific provisions described in sections 5.2.1.2 – 5.2.3.5, the surface linings of walls and ceilings should meet the classifications in table 7.

Table 7 Classification of linings		
Location	National class[1]	European Class[1][3][4]
Small rooms[2] of area not more than 30m²	3	D-s3, d2
Other rooms[2]	1	C-s3, d2
Other circulation spaces	0	B-s3, d2

Notes:

1. See section 5.1.3.5.

2. For meaning of room, see definition in appendix F.

3. The National classifications do not automatically equate with the equivalent classifications in the European column, therefore, products cannot typically assume a European class, unless they have been tested accordingly.

4. When a classification includes 's3, d2', this means that there is no limit set for smoke production and/or flaming droplets/particles.

5.2.1.2 Definition of walls

For the purpose of the performance of wall linings, a wall includes:

- the surface of glazing (except glazing in doors); and

- any part of a ceiling which slopes at an angle of more than 70° to the horizontal.

But a wall does not include:

- doors and door frames;

- window frames and frames in which glazing is fitted;

- architraves, cover moulds, picture rails, skirtings and similar narrow members; or

- fireplace surrounds, mantle shelves and fitted furniture.

5.2.1.3   Definition of ceilings

For the purposes of the performance of ceiling linings, a ceiling includes:

*   the surface of glazing;
*   any part of a wall which slopes at an angle of 70° or less to the horizontal;
*   the underside of a mezzanine or gallery; and
*   the underside of a roof exposed to the room below.

But a ceiling does not include:

*   trap doors and their frames;
*   the frames of windows or rooflights (see appendix F) and frames in which glazing is fitted; or
*   architraves, cover moulds, picture rails, exposed beams and similar narrow members.

5.2.2   Variations and special provisions

5.2.2.1   Walls

Parts of walls in rooms may be of a poorer performance than specified in section 5.2.1.1 and table 7 (but not poorer than Class 3 (National class) or Class D-s3, d2 (European class)), provided the total area of those parts in any one room does not exceed one half of the floor area of the room; and subject to a maximum of 60m^2.

Notice boards with a surface spread of flame are permitted in classrooms, but these should not extend more than 2.5m without having a break between them of not less than 0.4m, and should be located away from potential sources of ignition. Notice boards should not be provided in dead end corridors unless covered by a suitable material (eg, glass or polycarbonate), see section 3.1.6.

5.2.2.2   Fire-protecting suspended ceilings

A suspended ceiling can contribute to the overall fire resistance of a floor/ceiling assembly. Such a ceiling should satisfy section 5.2.1.1 and table 7. It should also meet the provisions of appendix A, table A3.

5.2.2.3   Fire-resisting ceilings

Cavity barriers are needed in some concealed floor or roof spaces (see section 6.4); however, this need can be reduced by the use of a fire-resisting ceiling below the cavity. Such a ceiling should comply with figure 32.

5.2.2.4   Rooflights

Rooflights should meet the relevant classification in section 5.2.1.1 and table 7. However plastic rooflights with at least a Class 3 rating may be used where section 5.2.1.1 calls for a higher standard, provided the limitations in table 8 and table 17 are observed.

Note: No guidance is currently possible on the performance requirements in the European fire tests as there is no generally accepted test and classification procedure.

5.2.2.5   Special applications

Any flexible membrane covering a structure (other than an air supported structure) should comply with the recommendations given in appendix A of BS 7157:1989.

Guidance on the use of PTFE-based materials for tension-membrane roofs and structures is given in a BRE report *Fire safety of PTFE-based materials used in buildings (BR 274, BRE 1994).*

5.2.3   Thermoplastic materials

5.2.3.1   General

Thermoplastic materials (see appendix A) which cannot meet the performance given in table 7, can nevertheless be used in windows, rooflights and lighting diffusers in suspended ceilings if they comply with the provisions described in sections 5.2.3.2 – 5.2.3.4. Flexible thermoplastic material may be used in panels to form a suspended ceiling if it complies with the guidance in section 5.2.3.5. The classifications used in sections 5.2.3.2 – 5.2.3.5, table 8 and figure 24 are explained in appendix A.

Note: No guidance is currently possible on the performance requirements in the European fire tests as there is no generally accepted test and classification procedure.

5.2.3.2 Windows and internal glazing

External windows to rooms (though not to circulation spaces) may be glazed with thermoplastic materials, if the material can be classified as a TP(a) rigid product.

Internal glazing should meet the provisions in section 5.2.1.1 and table 7.

Note 1: A 'wall' does not include glazing in a door (see section 5.2.1.2).

Note 2: Attention is drawn to the guidance on the safety of glazing in Approved Document N *Glazing – safety in relation to impact, opening and cleaning.*

5.2.3.3 Rooflights

Rooflights to rooms and circulation spaces (with the exception of protected stairways) may be constructed of a thermoplastic material if:

a. the lower surface has a TP(a) (rigid) or TP(b) classification;

b. the size and disposition of the rooflights accords with the limits in table 8 and with the guidance to B4 in tables 16 and 17.

5.2.3.4 Lighting diffusers

The following provisions apply to lighting diffusers which form part of a ceiling and are not concerned with diffusers of light fittings which are attached to the soffit of, or suspended beneath, a ceiling (see figure 25).

Lighting diffusers are translucent or open-structured elements that allow light to pass through. They may be part of a luminaire or used below rooflights or other sources of light.

Table 8 Limitations applied to thermoplastic rooflights and lighting diffusers in suspended ceilings and Class 3 plastic rooflights

Minimum classification of lower surface	Use of space below the diffusers or rooflight	Maximum area of each diffuser panel or rooflight[1]	Max total area of diffuser panels and rooflights as percentage of floor area of the space in which the ceiling is located	Minimum separation distance between diffuser panels or rooflights[1]
		(m²)	(%)	(m)
TP(a)	Any except protected stairway	No limit[2]	No limit	No limit
Class 3[3] or TP(b)	Rooms	5	50[4][5]	3[5]
	Circulation spaces except protected stairways	5	5[4]	3

Notes:

1. Smaller panels can be grouped together provided that the overall size of the group and the space between one group and any others satisfies the dimensions shown in figure 24.

2. Lighting diffusers of TP(a) flexible rating should be restricted to panels of not more than 5m² each, see section 5.2.3.5.

3. There are no limits on Class 3 material in small rooms. See section 5.2.1.1, table 7.

4. The minimum 3m separation specified in figure 24 between each 5m² must be maintained. Therefore, in some cases it may not also be possible to use the maximum percentage quoted.

5. Class 3 rooflights to rooms in lindustrial and other non-residential purpose groups may be spaced 1800mm apart provided the rooflights are evenly distributed and do not exceed 20% of the area of the room.

Figure 24 Layout restrictions on Class 3 plastic rooflights, TP(b) rooflights and TP(b) lighting diffusers

See table 8

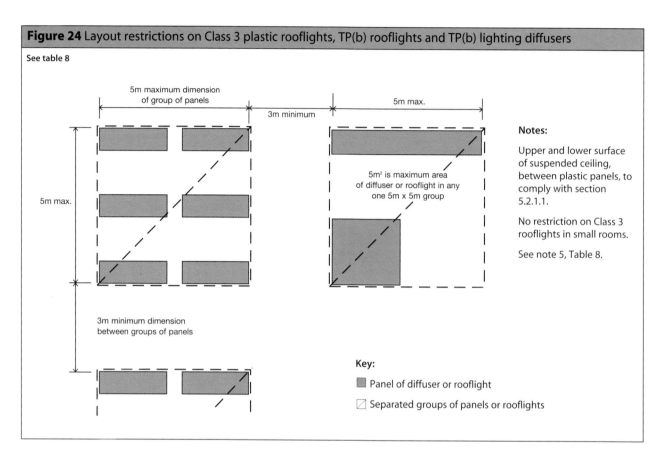

Notes:

Upper and lower surface of suspended ceiling, between plastic panels, to comply with section 5.2.1.1.

No restriction on Class 3 rooflights in small rooms.

See note 5, Table 8.

5m maximum dimension of group of panels

3m minimum

5m max.

5m max.

5m² is maximum area of diffuser or rooflight in any one 5m x 5m group

3m minimum dimension between groups of panels

Key:

▪ Panel of diffuser or rooflight

▱ Separated groups of panels or rooflights

Figure 25 Lighting diffuser in relation to ceiling

See section 5.2.3.4

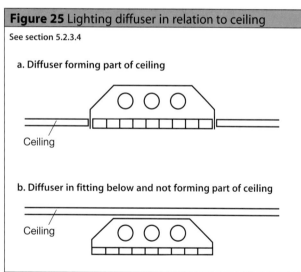

a. Diffuser forming part of ceiling

Ceiling

b. Diffuser in fitting below and not forming part of ceiling

Ceiling

Thermoplastic lighting diffusers should not be used in fire-protecting or fire-resisting ceilings, unless they have been satisfactorily tested as part of the ceiling system that is to be used to provide the appropriate fire protection.

Subject to the above paragraphs, ceilings to rooms and circulation spaces (but not protected stairways) may incorporate thermoplastic lighting diffusers if the following provisions are observed:

a. Wall and ceiling surfaces exposed within the space above the suspended ceiling (other than the upper surfaces of the thermoplastic panels) should comply with the general provisions of section 5.2.1.1 and table 7, according to the type of space below the suspended ceiling.

b. If the diffusers are of classification TP(a) (rigid), there are no restrictions on their extent.

c. If the diffusers are of classification TP(b), they should be limited in extent as indicated in table 8 and figure 24.

5.2.3.5 Suspended or stretched-skin ceilings

The ceiling of a room may be constructed either as a suspended or as a stretched skin membrane from panels of a thermoplastic material of the TP(a) flexible classification, provided that it is not part of a fire-resisting ceiling. Each panel should not exceed 5m² in area and should be supported on all its sides.

Agreement with the Building Control Body should be sought if it is intended to use a continuous sheet membrane roof over an area of low fire risk, eg, a swimming pool.

Section 6
Internal fire spread (structure)

6.1 Overview

6.1.1 Requirement B3 of the Building Regulations

(1) The building shall be designed and constructed so that, in the event of fire, its stability will be maintained for a reasonable period.

(2) A wall common to two or more buildings shall be designed and constructed so that it adequately resists the spread of fire between those buildings.

(3) Where reasonably necessary to inhibit the spread of fire within the building, measures shall be taken, to an extent appropriate to the size and intended use of the building, comprising either or both of the following:

   (a) sub-division of the building with fire-resisting construction; and/or

   (b) installation of suitable automatic fire suppression systems.

(4) The building shall be designed and constructed so that the unseen spread of fire and smoke within concealed spaces in its structure and fabric is inhibited.

6.1.2 Performance

Requirements of B3 will be met:

a. if the loadbearing elements of structure of the building are capable of withstanding the effects of fire for an appropriate period without loss of stability;

b. if the building is sub-divided by elements of fire-resisting construction into compartments;

c. if any openings in fire-separating elements (see appendix F) are suitably protected in order to maintain the integrity of the element (ie, the continuity of the fire separation); and

d. if any hidden voids in the construction are sealed and sub-divided to inhibit the unseen spread of fire and products of combustion, in order to reduce the risk of structural failure and the spread of fire, in so far as they pose a threat to the safety of people in and around the building.

The extent to which any of these measures are necessary is dependent on the use of the building and, in some cases, its size and on the location of the element of construction.

Protection of the property and contents against fire may require the use of fire-resisting glazing with an insulation performance (rather than integrity only) because of the risk of exposure of the fabric of the building to prolonged periods of post-flashover fire conditions.

6.1.3 Introduction

Guidance on loadbearing elements of structure is given in section 6.2. Section 6.3 is concerned with the sub-division of a building into compartments and section 6.4 makes provisions about concealed spaces (or cavities). Section 6.5 gives information on the protection of openings and on fire-stopping. Common to all these sections and to other provisions of Part B, is the property of fire resistance.

6.1.3.1 Fire resistance

The fire resistance of an element of construction is a measure of its ability to withstand the effects of fire in one or more ways, as follows:

- resistance to collapse, ie, the ability to maintain loadbearing capacity (which applies to loadbearing elements only) (R);

- resistance to fire penetration, ie, an ability to maintain the integrity of the element (E); and

- resistance to the transfer of excessive heat, ie, an ability to provide insulation from high temperatures (I).

'Elements of structure' is the term applied to the main structural loadbearing elements, such as structural frames, floors and loadbearing walls. Compartment walls are treated as elements of structure although they are not necessarily loadbearing. Roofs, unless they serve the function of a floor, are not treated as elements of structure. External walls, such as curtain walls or other forms of cladding which transmit only self weight and wind loads and do not transmit floor load, are not regarded as loadbearing for the purposes of this section, although they may need fire resistance to satisfy requirement B4 (see sections 7.2 and 7.3).

Loadbearing elements may or may not have a fire-separating function. Similarly, fire-separating elements may or may not be loadbearing.

6.1.3.2 Guidance elsewhere in this guide concerning fire resistance

There is guidance in sections 4.3 – 4.5 concerning the use of fire-resisting construction to protect means of escape. There is guidance in section 7.2 about fire resistance of external walls to restrict the spread of fire between buildings. There is guidance in section 8.4 about fire resistance in the construction of fire-fighting shafts. Appendix A gives information on methods of test and performance for elements of construction. Appendix C gives information on fire doors. Appendix E gives information on methods of measurement. Appendix F gives definitions.

6.2 Loadbearing elements of structure

6.2.1 Introduction

Premature failure of the structure can be prevented by provisions for loadbearing elements of structure to have a minimum standard of fire resistance, in terms of resistance to collapse or failure of loadbearing capacity. The purpose in providing the structure with fire resistance is threefold, namely:

- to minimise the risk to the occupants, some of whom may have to remain in the building for some time while evacuation proceeds if the building is a large one;
- to reduce the risk to firefighters, who may be engaged on search or rescue operations;
- to reduce the danger to people in the vicinity of the building, who might be hurt by falling debris or as a result of the impact of the collapsing structure on other buildings; and
- to provide structural resilience against fire and limit the possibility of fire spread.

6.2.2 Fire resistance standard

Elements of structure such as structural frames, beams, columns, loadbearing walls (internal and external), floor structures and gallery structures, should have at least the fire resistance given in appendix A, table A1.

6.2.2.1 Application of the fire resistance standards for loadbearing elements

The measures set out in appendix A include provisions to ensure that where one element of structure supports or gives stability to another element of structure, the supporting element has no less fire resistance than the other element (see notes to table A2). The measures also provide for elements of structure that are common to more than one building or compartment, to be constructed to the standard of the greater of the relevant provisions. Special provisions about fire resistance of elements of structure in single storey buildings are also given and there are concessions in respect of fire resistance of elements of structure in basements where at least one side of the basement is open at ground level.

6.2.2.2 Exclusions from the provisions for elements of structure

The following are excluded from the definition of element of structure for the purposes of these provisions:

a. a structure that only supports a roof, unless:

   i. the roof performs the function of a floor, such as for parking vehicles, or as a means of escape; or

   ii. the structure is essential for the stability of an external wall which needs to have fire resistance;

b. the lowest floor of the building;

c. a platform floor; and

d. a loading gallery, fly gallery, stage grid, lighting bridge, or any gallery provided for similar purposes or for maintenance and repair (see definition of 'Element of structure' in appendix F).

6.2.2.3 Additional guidance

Guidance in other sections of this guide may also apply if a loadbearing wall is:

- a compartment wall (this includes a wall common to two buildings);
- a wall enclosing a place of special fire hazard;
- protecting a means of escape;
- an external wall; or
- enclosing a fire-fighting shaft.

If a floor is also a compartment floor, see section 6.3.

The ASFP Yellow Book publication *Fire protection of structural steel in buildings – 4th Edition*, ISBN 1 870409 221, provides up to date advice on all aspects of fire protection to steelwork, including new information on cellular steel beams. It is available for free download from www.asfp.org.uk

6.3 Compartmentation

6.3.1 Introduction

When a building is constructed to prevent the spread of fire from another part of the same building or an adjoining building, it is said to be compartmented. The compartments may consist of single or multiple rooms, spaces or storeys. The compartments should be constructed so that their relevant boundaries are fire-resisting.

The purpose of compartmentation is to contain fire and limit its extent to the place of origin. This is achieved through design, construction and product selection to:

- provide escape and access routes which are protected from the fire and its effects, that is flames, heat, and hot gases;
- minimise the impact of fire on the fabric of the building; and
- contain fire for more effective action by suppression systems and fire fighters.

The fundamental foundation for fire safety should be achieved by the building itself. Where necessary and appropriate this requires the use of products which are specifically designed and tested for their resilience against fire.

With the emphasis on new build, extensions and refurbishments that allow greater flexibility of use, the simple option of building a series of fire-tight boxes in school buildings may be considerably reduced. However, with respect to the acoustic requirements of the building to provide 'suitable indoor ambient noise levels for clear communication of speech between teacher and pupil, between pupils and for study activities' (as dealt with in AD E and BB 93) these will result in fire-tight enclosures within the overall design. The distribution of barriers to restrict fire spread will involve a greater degree of thought and recognition of the risks posed by the layout to the users and the occupants. A balance will need to be struck between the everyday needs in the building and security and fire safety.

Previous guidance Building Bulletin 7 *Fire and the design of education buildings* (ref), stressed the importance of mainly cellular room layouts in order to contain any fires. Whilst this remains valid, fire safety can be achieved to a similar level by utilising larger, more open spaces although that puts a greater onus on the remaining elements to perform well in the event of fire. This means that durability in service and resistance to abuse is more critical and must be considered when specifying in such areas of constructional weaknesses as junctions of wall to wall, wall to ceiling, etc. Section 3.8 includes guidance on robust constructions that can be used for fire separating walls and floors to withstand mechanical damage and reduce the effects of a fire.

The spread of fire within a building can be restricted by sub-dividing it into compartments separated from one another by walls and/or floors of fire-resisting construction. The object is fourfold:

- to prevent rapid fire spread which could trap occupants of the building;
- to reduce the chance of fires becoming large, on the basis that large fires are more dangerous, not only to occupants and Fire and Rescue Service personnel, but also to people in the vicinity of the building;

- to provide protected zones linked by protected escape ways to a place of safety outside the building; and

- to reinforce the stability and resilience of the building fabric against damage by fire so that it may resist the longer term impact of fire once the occupants have left the building.

Compartmentation is complementary to provisions made in sections 4.3 – 4.5 for the protection of escape routes and to provisions made in sections 7.2 – 7.4 against the spread of fire between buildings.

Compartmentation is also beneficial in reducing the extent of fire damage to a building. Other than for small premises, insurers may require a degree of compartmentation such that the maximum loss from a fire should never exceed 25% of the whole building.

The appropriate degree of sub-division depends on:

- the use of and fire load in the building, which affects the potential for fires and the severity of fires, as well as the ease of evacuation;

- the height to the floor of the top storey in the building, which is an indication of the ease of evacuation and the ability of the Fire and Rescue Service to intervene effectively; and

- the availability of a sprinkler system which affects the growth rate of the fire and may suppress it altogether.

Sub-division is achieved using compartment walls and compartment floors. The circumstances in which they are needed are given in sections 6.3.2.1 – 6.3.2.3.

Provisions for the construction of compartment walls and compartment floors are given in sections 6.3.3.1 onwards. These construction provisions vary according to the function of the wall or floor.

6.3.1.1　Special forms of compartmentation

Special forms of compartmentation to which particular construction provisions apply, are:

a. walls common to two or more buildings, see section 6.3.2.1;

b. walls dividing buildings into separated parts, see section 6.3.2.1; and

c. construction enclosing places of special fire hazard, see section 6.3.2.2.

6.3.1.2　Junctions

For compartmentation to be effective, there should be continuity at the junctions of the fire-resisting elements enclosing a compartment and any openings from one compartment to another should not present a weakness.

Excessive deflections should be prevented, as these will tend to jeopardise the requirement for compartment integrity. Large deflections will also make it more difficult to successfully carry out building refurbishment following a fire. For further recommendations on dealing with deflections, see appendix 3D of the *LPC Design Guide for the Fire Protection of Buildings* (ref).

6.3.1.3　Protected shafts

Spaces that connect compartments, such as stairways and service shafts, need to be protected to restrict fire spread between the compartments and they are termed protected shafts. Any walls or floors bounding a protected shaft are considered to be compartment walls or floors.

6.3.1.4　Buildings containing one or more atria

Detailed advice on all issues relating to the incorporation of atria in buildings is given in BS 5588-7:1997. However, it should be noted that for life safety purposes, the standard is relevant only where the atrium breaches any compartmentation.

A fire engineering solution should be used whenever unusual materials or designs are used, eg, ETFE for roofs. See also *The design and protection of new school buildings and sites*, 2005 (ref).

6.3.2 Provision of compartmentation

The maximum dimensions of compartments within schools are given in table 9.

Table 9 Maximum dimensions of compartments within schools	
Floor area of any one storey in the school or any one storey in a compartment (m²)	
In multi-storey schools	In single storey schools
Not sprinklered – 800	Not sprinklered – 800
Sprinklered – 2000	Sprinklered – No limit

Note: 'Sprinklered' means that the school is fitted throughout with an automatic system meeting the relevant recommendations of the SSLD on Spinklers in Schools (ref) plus requirements for 'life safety'. The benefit to the designer from the installation of such a suppression system is that compartments can more than double in size.

6.3.2.1 General

Compartment walls and compartment floors should be provided in the circumstances described below, with the proviso that the lowest floor in a building does not need to be constructed as a compartment floor, provided there is no basement. Sections 6.3.2.1 – 6.3.2.3 give guidance on the provision of compartmentation in different building types. Information on the construction of compartment walls and compartment floors in different circumstances is given in sections 6.3.3.1 – 6.3.3.6. Provisions for the protection of openings in compartment walls and compartment floors are given in sections 6.3.4.1 – 6.3.4.3.

A wall common to two or more buildings should be constructed as a compartment wall.

Parts of a building that are occupied mainly for different purposes should be separated from one another by compartment walls and/or compartment floors. This does not apply where one of the different purposes is ancillary to the other.

In order to reduce the extent of property damage, all floors in unsprinklered schools should be compartment floors.

6.3.2.2 Places of special fire risk

Every place of special fire hazard (see section 3.1 and appendix F) should be enclosed with fire-resisting construction; see table A1, item 12, in appendix A.

Consideration should also be given to the fire protection of parts of the building housing valuable assets, in order to reduce property damage and/or business interruption following a fire.

Parts of the school used out of normal hours by school clubs or members of the public could be considered as higher risk, by virtue of the fact that they are in use for longer periods of the day. If these parts form separate compartments, the benefits for property protection are twofold:

- there is less chance that a fire starting in the area used outside normal school hours will spread to the rest of the school; and

- there is less chance that a fire starting elsewhere in the school will spread to affect the areas of community use.

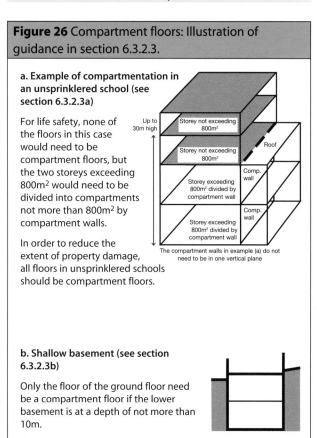

Figure 26 Compartment floors: Illustration of guidance in section 6.3.2.3.

a. Example of compartmentation in an unsprinklered school (see section 6.3.2.3a)

For life safety, none of the floors in this case would need to be compartment floors, but the two storeys exceeding 800m² would need to be divided into compartments not more than 800m² by compartment walls.

In order to reduce the extent of property damage, all floors in unsprinklered schools should be compartment floors.

Up to 30m high

Storey not exceeding 800m²
Storey not exceeding 800m²
Roof
Storey exceeding 800m² divided by compartment wall
Comp. wall
Storey exceeding 800m² divided by compartment wall
Comp. wall

The compartment walls in example (a) do not need to be in one vertical plane

b. Shallow basement (see section 6.3.2.3b)

Only the floor of the ground floor need be a compartment floor if the lower basement is at a depth of not more than 10m.

Note: Any such walls and floors are not compartment walls and compartment floors.

6.3.2.3   General

The following walls and floors should be constructed as compartment walls and compartment floors:

- every wall needed to sub-divide the building to observe the size limits on compartments given in table 9 (see figure 26a); and
- the floor of the ground storey if the building has one or more basements (see figure 26b).

6.3.3   Construction of compartment walls and compartment floors

6.3.3.1   General

Every compartment wall and compartment floor should:

- form a complete barrier to fire between the compartments they separate; and
- have the appropriate fire resistance as indicated in appendix A, tables A1 and A2.

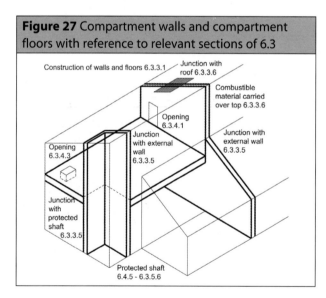

Figure 27 Compartment walls and compartment floors with reference to relevant sections of 6.3

Construction of walls and floors 6.3.3.1
Junction with roof 6.3.3.6
Combustible material carried over top 6.3.3.6
Opening 6.3.4.1
Junction with external wall 6.3.3.5
Junction with external wall 6.3.3.5
Opening 6.3.4.3
Junction with protected shaft 6.3.3.5
Protected shaft 6.4.5 - 6.3.5.6

Note 1: Timber beams, joists, purlins and rafters may be built into or carried through a masonry or concrete compartment wall if the openings for them are kept as small as practicable and then fire-stopped. If trussed rafters bridge the wall, they should be designed so that failure of any part of the truss due to a fire in one compartment will not cause failure of any part of the truss in another compartment.

Note 2: Where services are incorporated within the construction that could provide a potential source of ignition, care should be taken to ensure the risk of fire developing and spreading

prematurely into adjacent compartments is controlled.

Additional recommendations for property protection and the minimisation of disruption are that compartment walls and floors should be constructed of robust materials (see section 3.8) to withstand accidental and mechanical damage, and that in the event of a fire they should be capable of being reinstated with minimal disturbance to adjoining buildings.

6.3.3.2   Compartment walls between buildings

Compartment walls that are common to two or more buildings should run the full height of the building in a continuous vertical plane. Thus adjoining buildings should only be separated by walls, not floors.

6.3.3.3   Separated parts of buildings

Compartment walls used to form a separated part of a building (so that the separated parts can be assessed independently for the purpose of determining the appropriate standard of fire resistance) should run the full height of the building in a continuous vertical plane. The two separated parts may have different standards of fire resistance.

6.3.3.4   Other compartment walls

Compartment walls not described in sections 6.3.3.2 and 6.3.3.3 should run the full height of the storey in which they are situated.

Compartment walls in a top storey beneath a roof should be continued through the roof space (see definition of compartment in appendix F).

6.3.3.5   Junction of compartment wall or compartment floor with other walls

Where a compartment wall or compartment floor meets another compartment wall or an external wall, the junction should maintain the fire resistance of the compartmentation. Fire-stopping should meet the provisions of section 6.5.4.

At the junction of a compartment floor with an external wall that has no fire resistance (such as a curtain wall) the external wall should be restrained at floor level to reduce the movement of the wall away from the floor when exposed to fire.

It is important that fire-resistant fire stopping is securely fixed between the floor slab and the external wall.

Compartment walls should be able to accommodate the predicted deflection of the floor above by either:

• having a suitable head detail between the wall and the floor, that can deform but maintain integrity when exposed to a fire; or

• the wall may be designed to resist the additional vertical load from the floor above as it sags under fire conditions and thus maintain integrity.

Note: Where compartment walls are located within the middle half of a floor between vertical supports, the predicted deflection may be assumed to be 40mm unless a smaller value can be justified by assessment. Outside this area the limit can be reduced linearly to zero at the supports. For steel beams that do not have the required fire resistance, reference should be made to SCI Publication 288 *Fire safe design: A new approach to multi-storey steel-framed buildings* (Second Edition) 2000 (ISBN: 1 85942 169 5).

6.3.3.6 Junction of compartment wall with roof

A compartment wall should be taken up to meet the underside of the roof covering or deck, with fire-stopping where necessary at the wall/roof junction to maintain the continuity of fire resistance. The compartment wall should also be continued across any eaves cavity (see section 6.3.3.1).

If a fire penetrates a roof near a compartment wall there is a risk that it will spread over the roof to the adjoining compartment. To reduce this risk, in buildings not more than 15m high, a zone of the roof 1500mm wide on either side of the wall should have a covering of designation AA, AB or AC (see appendix A) on a substrate or deck of a material of limited combustibility, as set out in figure 28a.

Note 1: Thermoplastic rooflights which, by virtue of section 7.4.2.2, are regarded as having an AA (National class) designation or $B_{ROOF}(t4)$ (European class) classification are not suitable for use in the zone described above.

Note 2: Double-skinned insulated roof sheeting, with a thermoplastic core, should incorporate a band of material of limited combustibility at least 300mm wide centred over the wall.

In buildings not more than 15m high, combustible boarding used as a substrate to the roof covering, wood wool slabs, or timber tiling battens, may be carried over the compartment wall provided that they are fully bedded in mortar or other suitable material over the width of the wall (see figure 28b).

As an alternative to the above the compartment wall may be extended up through the roof for a height of at least 375mm above the top surface of the adjoining roof covering. Where there is a height difference of at least 375mm between two roofs or where the roof coverings on either side of the wall are AA, AB or AC this height may be reduced to 200mm (see figure 28c).

For the purposes of property protection, the compartment wall should be extended at least 500mm above the adjoining roof covering.

6.3.4 Openings in compartmentation

Glazed apertures in compartment walls shall be fire-resisting glazed systems. All fire-resisting glass types only function as intended when installed as part of a fire-resisting glazed system which includes matched components. The whole system shall be fire-resisting and shall have relevant evidence of fire performance provided in an appropriate fire test report. Approved fire-resisting glazed systems shall be installed as specified in the test report, and there should be no changes without appropriate authorisation from the glass manufacturer.

If the compartment is sprinklered then modified toughened fire-resisting glass types should not be used because of the risk of unpredictable shattering of the glass when the sprinklers activate in a fire.

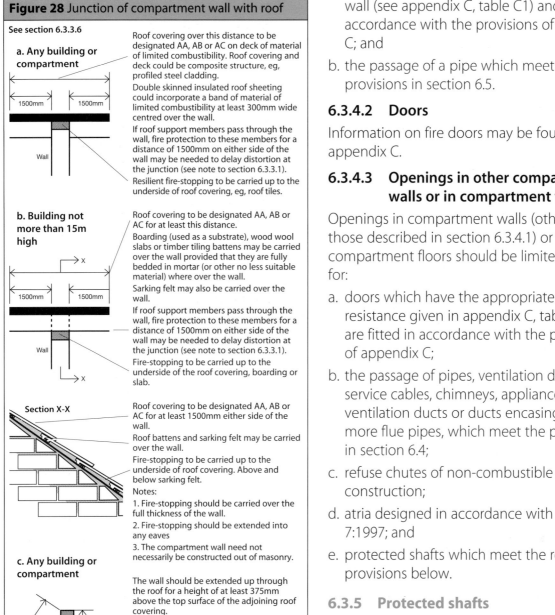

Figure 28 Junction of compartment wall with roof

See section 6.3.3.6

a. Any building or compartment

1500mm   1500mm

Wall

Roof covering over this distance to be designated AA, AB or AC on deck of material of limited combustibility. Roof covering and deck could be composite structure, eg, profiled steel cladding.

Double skinned insulated roof sheeting could incorporate a band of material of limited combustibility at least 300mm wide centred over the wall.

If roof support members pass through the wall, fire protection to these members for a distance of 1500mm on either side of the wall may be needed to delay distortion at the junction (see note to section 6.3.3.1).

Resilient fire-stopping to be carried up to the underside of roof covering, eg, roof tiles.

b. Building not more than 15m high

1500mm   1500mm

Wall

Roof covering to be designated AA, AB or AC for at least this distance.

Boarding (used as a substrate), wood wool slabs or timber tiling battens may be carried over the wall provided that they are fully bedded in mortar (or other no less suitable material) where over the wall.

Sarking felt may also be carried over the wall.

If roof support members pass through the wall, fire protection to these members for a distance of 1500mm on either side of the wall may be needed to delay distortion at the junction (see note to section 6.3.3.1).

Fire-stopping to be carried up to the underside of the roof covering, boarding or slab.

Section X-X

Wall

Roof covering to be designated AA, AB or AC for at least 1500mm either side of the wall.

Roof battens and sarking felt may be carried over the wall.

Fire-stopping to be carried up to the underside of roof covering. Above and below sarking felt.

Notes:

1. Fire-stopping should be carried over the full thickness of the wall.

2. Fire-stopping should be extended into any eaves

3. The compartment wall need not necessarily be constructed out of masonry.

c. Any building or compartment

At least 375mm   At least 375mm

Roof covering

Roof covering

Wall

The wall should be extended up through the roof for a height of at least 375mm above the top surface of the adjoining roof covering.

Where there is a height difference of at least 375mm between two roofs or where the roof coverings on either side of the wall are AA, AB or AC the height of the upstand/parapet wall above the highest roof may be reduced to 200mm.

6.3.4.1 Openings in compartment walls separating buildings or occupancies

Any openings in a compartment wall which is common to two or more buildings, or between different occupancies in the same building, should be limited to those for:

a. a door which is needed to provide a means of escape in case of fire and which has the same fire resistance as that required for the

wall (see appendix C, table C1) and is fitted in accordance with the provisions of appendix C; and

b. the passage of a pipe which meets the provisions in section 6.5.

6.3.4.2 Doors

Information on fire doors may be found in appendix C.

6.3.4.3 Openings in other compartment walls or in compartment floors

Openings in compartment walls (other than those described in section 6.3.4.1) or compartment floors should be limited to those for:

a. doors which have the appropriate fire resistance given in appendix C, table C1 and are fitted in accordance with the provisions of appendix C;

b. the passage of pipes, ventilation ducts, service cables, chimneys, appliance ventilation ducts or ducts encasing one or more flue pipes, which meet the provisions in section 6.4;

c. refuse chutes of non-combustible construction;

d. atria designed in accordance with BS 5588-7:1997; and

e. protected shafts which meet the relevant provisions below.

6.3.5 Protected shafts

Any stairway or other shaft passing directly from one compartment to another should be enclosed in a protected shaft so as to delay or prevent the spread of fire between compartments.

There are additional provisions in sections 4.3 – 4.5 for protected shafts that are protected stairways and in section 8.4 if the stairway also serves as a fire-fighting stair.

6.3.5.1 Uses for protected shafts

The uses of protected shafts should be restricted to stairs, lifts, escalators, chutes, ducts and pipes. Sanitary accommodation and washrooms may be included in protected shafts.

6.3.5.2 Construction of protected shafts

The construction enclosing a protected shaft (see figure 29) should:

a. form a complete barrier to fire between the different compartments which the shaft connects;

b. have the appropriate fire resistance given in appendix A, table A1, except for uninsulated glazed screens which meet the provisions of section 6.3.5.3; and

c. satisfy the provisions about their ventilation and the treatment of openings in section 6.3.5.5 – 6.3.5.6.

6.3.5.3 Uninsulated glazed screens to protected shafts

If the conditions given below are satisfied, an uninsulated glazed screen may be incorporated in the enclosure to a protected shaft between a stair and a lobby or corridor which is entered from the stair. The conditions to be satisfied are:

a. the recommended standard of fire resistance for the stair enclosure is not more than 60 minutes; and

b. the glazed screen:

 i. has at least 30 minutes fire resistance in terms of integrity; and

 ii. meets the guidance in appendix A, table A4, on the limits on areas of uninsulated glazing; and

c. the lobby or corridor is enclosed to at least a 30 minute standard.

If the above measures to protect the lobby or corridor are not provided, the enclosing walls should comply with appendix A, table A1 (item 7c) and the doors with the guidance in appendix A, table A4.

6.3.5.4 Pipes for oil or gas and ventilation ducts in protected shafts

If a protected shaft contains a stair and/or a lift, it should not also contain a pipe conveying oil (other than in the mechanism of a hydraulic lift) or contain a ventilating duct (other than a duct provided for the purposes of pressurizing the stairway to keep it smoke free; or a duct provided solely for ventilating the stairway).

Any pipe carrying natural gas or LPG in such a shaft should be of screwed steel or of all welded steel construction, installed in accordance with the *Pipelines Safety Regulations 1996*, SI 1996 No 825 and the *Gas Safety (Installation and use) Regulations 1998*, SI 1998 No 2451.

Note: A pipe is not considered to be contained within a protected shaft if the pipe is completely separated from that protected shaft by fire-resisting construction.

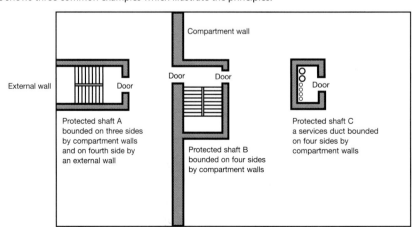

Figure 29 Protected Shafts

See section 6.3.5 – 6.5.3.2

Protected shafts provide for the movement of people (eg, stairs, lifts), or for passage of goods, air or services such as pipe or cables between different compartments. The elements enclosing the shaft (unless formed by adjacent external walls) are compartment walls and floors. The figure shows three common examples which illustrate the principles.

Compartment wall

External wall   Door   Door   Door

Protected shaft A
bounded on three sides
by compartment walls
and on fourth side by
an external wall

Protected shaft B
bounded on four sides
by compartment walls

Protected shaft C
a services duct bounded
on four sides by
compartment walls

The shaft structure (including any openings) should meet the relevant provisions for: compartment walls (see sections 6.3.3.1 to 6.3.4.3), external walls (see sections 7.2 and 7.3 and figure 22).

6.3.5.5 Ventilation of protected shafts conveying gas

A protected shaft conveying piped flammable gas should be adequately ventilated direct to the outside air by ventilation openings at high and low level in the shaft.

Any extension of the storey floor into the shaft should not compromise the free movement of air over the entire length of the shaft. Guidance on such shafts, including sizing of the ventilation openings, is given in BS 8313:1997.

6.3.5.6 Openings into protected shafts

Generally an external wall of a protected shaft does not need to have fire resistance.

However, there are some provisions for fire resistance of external walls of fire-fighting shafts in BS 5588-5:2004, which is the relevant guidance called up by sections 8.4.4 (also see section 17.13 of AD B) and of external walls to protected stairways (which may also be protected shafts) in section 4.4.6.7.

Openings in other parts of the enclosure to a protected shaft should be limited as follows:

a. Where part of the enclosure to a protected shaft is a wall common to two or more buildings, only the following openings should be made in that wall:

   i. a door which is needed to provide a means of escape in case of fire; and which has the same fire resistance as that required for the wall (see appendix C, table C1); and is fitted in accordance with the provisions of appendix C; and/or

   ii. the passage of a pipe which meets the provisions in section 6.5.

b. Other parts of the enclosure (other than an external wall) should only have openings for:

   i. doors which have the appropriate fire resistance given in appendix C, table C1 and are fitted in accordance with the provisions of appendix C;

   ii. the passage of pipes which meet the provisions in section 6.5;

   iii. inlets to, outlets from and openings for a ventilation duct, (if the shaft contains or serves as a ventilating duct) which meet the provisions in section 6.5; and/or

   iv. the passage of lift cables into any lift machine room. If the machine room is at the bottom of the shaft, the openings should be as small as practicable.

6.4 Concealed spaces (cavities)

6.4.1 Introduction

Hidden routes for fire spread include concealed spaces above suspended ceilings. Vertical fire separation must continue up to the underside of the true ceiling thus ensuring compartmentation is achieved.

False ceilings are often the route for electrical and other links within a school and care should be taken to ensure that their installation is checked to avoid breaching compartmentation and that there are no faults and the possibility of fires that may remain undetected for a period.

Concealed spaces or cavities in the construction of a building provide a ready route for smoke and flame spread. This is particularly so in the case of voids in, above and below the construction of a building, eg, walls, floors, ceilings and roofs. As any spread is concealed, it presents a greater danger than would a more obvious weakness in the fabric of the building.

6.4.2 Provision of cavity barriers

Provisions for cavity barriers are given below for specified locations. The provisions necessary to restrict the spread of smoke and flames through cavities are broadly for the purpose of sub-dividing:

a. cavities, which could otherwise form a pathway around a fire-separating element and closing the edges of cavities; therefore reducing the potential for unseen fire spread. **Note:** These should not be confused with fire- stopping details, see section 6.5 and figure 30 (see also sections 6.4.3.1 – 6.4.3.4); and

b. extensive cavities (see sections 6.4.4 – 6.4.4.1).

Consideration should also be given to the construction and fixing of cavity barriers provided for these purposes and the extent to which openings in them should be protected. For guidance on these issues, see section 6.4.5.

Figure 30 Provisions for cavity barriers

See section 6.3.5 – 6.5.3.2

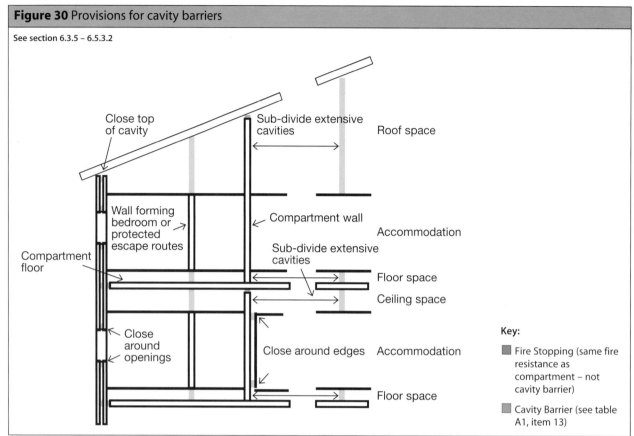

Close top of cavity

Sub-divide extensive cavities

Roof space

Wall forming bedroom or protected escape routes

Compartment wall

Accommodation

Sub-divide extensive cavities

Compartment floor

Floor space

Ceiling space

Close around openings

Close around edges   Accommodation

Floor space

Key:

■ Fire Stopping (same fire resistance as compartment – not cavity barrier)

▨ Cavity Barrier (see table A1, item 13)

6.4.3 Pathways around fire-separating elements

6.4.3.1 Junctions and cavity closures

Cavity barriers should be provided to close the edges of cavities, including around openings.

Cavity barriers should also be provided:

- at the junction between an external cavity wall (except where the cavity wall complies with figure 31) and every compartment floor and compartment wall; and

- at the junction between an internal cavity wall (except where the cavity wall complies with figure 31) and every compartment floor, compartment wall, or other wall or door assembly which forms a fire-resisting barrier.

It is important to continue any compartment wall up through a ceiling or roof cavity to maintain the standard of fire resistance – therefore compartment walls should be carried up full storey height to a compartment floor or to the roof as appropriate (see sections 6.3.3.2 – 6.3.3.4). It is therefore not appropriate to complete a line of compartmentation by fitting cavity barriers above them.

Figure 31 Cavity wall excluded from provisions for cavity barriers

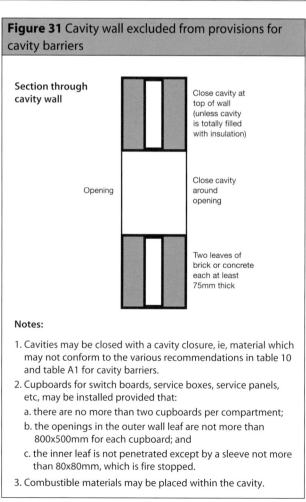

Section through cavity wall

Close cavity at top of wall (unless cavity is totally filled with insulation)

Opening

Close cavity around opening

Two leaves of brick or concrete each at least 75mm thick

Notes:

1. Cavities may be closed with a cavity closure, ie, material which may not conform to the various recommendations in table 10 and table A1 for cavity barriers.

2. Cupboards for switch boards, service boxes, service panels, etc, may be installed provided that:

   a. there are no more than two cupboards per compartment;

   b. the openings in the outer wall leaf are not more than 800x500mm for each cupboard; and

   c. the inner leaf is not penetrated except by a sleeve not more than 80x80mm, which is fire stopped.

3. Combustible materials may be placed within the cavity.

6.4.3.2 Protected escape routes

For a protected escape route, a cavity that exists above or below any fire-resisting construction because the construction is not carried to full storey height or, in the case of a top storey, to the underside of the roof covering, should either be:

a. fitted with cavity barriers on the line of the enclosure(s) to the protected escape route; or

b. for cavities above the fire-resisting construction, enclosed on the lower side by a fire-resisting ceiling which extends throughout the building, compartment or separated part (see figure 32).

6.4.3.3 Double-skinned corrugated or profiled roof sheeting

Cavity Barriers need not be provided between double-skinned corrugated or profiled insulated roof sheeting, if the sheeting is a material of limited combustibility and both surfaces of the insulating layer have a surface spread of flame of at least Class 0 or 1 (National class) or Class C-s3, d2 or better (European class) (see appendix A) and make contact with the inner and outer skins of cladding (see figure 33).

Note: See also section 6.3.3.6, note 2 regarding the junction of a compartment wall with a roof.

Note: When a classification includes 's3, d2', this means that there is no limit set for smoke production and/or flaming droplets/particles.

6.4.3.4 Cavities affecting alternative escape routes

Cavity barriers may be needed where corridors are sub-divided to prevent alternative escape routes being simultaneously affected by fire and/or smoke (see section 4.3.2.16 and figure 18).

6.4.4 Extensive cavities

Cavity barriers should be used to sub-divide any cavity, including any roof space, so that the distance between cavity barriers does not exceed the dimensions given in table 10.

6.4.4.1 Maximum dimensions of concealed spaces

Table 10 sets out maximum dimensions for undivided concealed spaces. With the exceptions given in the rest of this section, extensive concealed spaces should be sub-divided to comply with the dimensions in table 10.

The provisions in table 10 do not apply to any cavity described below:

a. in a wall which should be fire-resisting only because it is loadbearing;

b. in a masonry or concrete external cavity wall shown in figure 31;

c. in any floor or roof cavity above a fire-resisting ceiling, as shown in figure 32 and which extends throughout the building or compartment subject to a 30m limit on the extent of the cavity; or

Table 10 Maximum dimensions of cavities			
Location of cavity	Class of surface/product exposed in cavity (excluding the surface of any pipe, cable or conduit, or any insulation to any pipe)		Maximum dimensions in any direction (m)
	National class	European class	
Between a roof and a ceiling	Any	Any	20
Any other cavity	Class 0 or Class 1	Class A1 or Class A2-s3,d2 or Class B-s3,d2 or Class C-s3,d2	20
	Not Class 0 or Class 1	Not any of the above classes	10

Notes:

1. Exceptions to these provisions are given in section 6.4.4.1.

2. The national classifications do not automatically equate with the equivalent classifications in the European column, therefore products cannot typically assume a European class unless they have been tested accordingly.

3. When a classification includes 's3, d2', this means that there is no limit set for smoke production and/or flaming droplets/particles.

d. formed behind the external skin of an external cladding system with a masonry or concrete inner leaf at least 75mm thick, or by overcladding an existing masonry (or concrete) external wall, or an existing concrete roof, provided that the cavity does not contain combustible insulation; or

e. between double-skinned corrugated or profiled insulated roof sheeting, if the sheeting is a material of limited combustibility and both surfaces of the insulating layer have a surface spread of flame of at least Class 0 or 1 (National class) or Class C-s3, d2 or better (European class) (see appendix A) and make contact with the inner and outer skins of cladding (see figure 33); or

f. below a floor next to the ground or oversite concrete, if the cavity is less than 1000mm in height or if the cavity is not normally

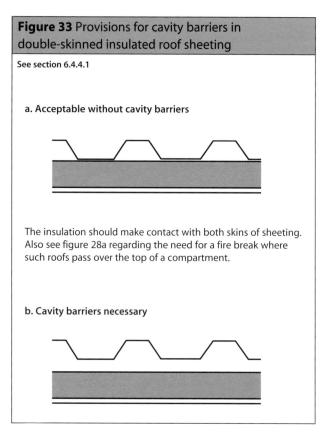

Figure 33 Provisions for cavity barriers in double-skinned insulated roof sheeting

See section 6.4.4.1

a. Acceptable without cavity barriers

The insulation should make contact with both skins of sheeting. Also see figure 28a regarding the need for a fire break where such roofs pass over the top of a compartment.

b. Cavity barriers necessary

accessible by persons, unless there are openings in the floor such that it is possible for combustibles to accumulate in the cavity (in which case cavity barriers should be provided and access should be provided to the cavity for cleaning).

Note: When a classification includes 's3, d2', this means that there is no limit set for smoke production and/or flaming droplets/particles.

Where any single room with a ceiling cavity or underfloor service void exceeds the dimensions given in table 10, cavity barriers need only be provided on the line of the enclosing walls/partitions of that room, subject to:

a. the cavity barriers being no more than 40m apart; and

b. the surface of the material/product exposed in the cavity being Class 0 or Class 1 (National class) or Class C-s3, d2 or better (European class).

Note: When a classification includes 's3, d2', this means that there is no limit set for smoke production and/or flaming droplets/particles.

Where the concealed space is an undivided area which exceeds 40m (this may be in both

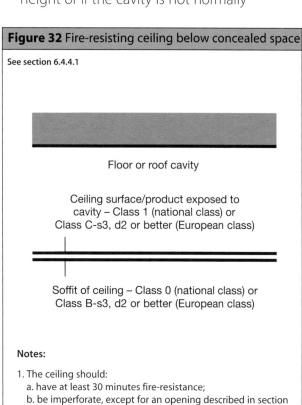

Figure 32 Fire-resisting ceiling below concealed space

See section 6.4.4.1

Floor or roof cavity

Ceiling surface/product exposed to cavity – Class 1 (national class) or Class C-s3, d2 or better (European class)

Soffit of ceiling – Class 0 (national class) or Class B-s3, d2 or better (European class)

Notes:

1. The ceiling should:
   a. have at least 30 minutes fire-resistance;
   b. be imperforate, except for an opening described in section 6.4.5;
   c. extend throughout the building or compartment; and
   d. not be easily demountable.

2. The National classifications do not automatically equate with the equivalent classifications in the European column, therefore products cannot typically assume a European class unless they have been tested accordingly.

3. When a classification includes 's3, d2', this means that there is no limit set for smoke production and/or flaming droplets/particles.

directions on plan) there is no limit to the size of the cavity if:

a. the room and the cavity together are compartmented from the rest of the building;

b. an automatic fire detection and alarm system meeting the relevant recommendations of BS 5839-1:2002 is fitted in the building. Detectors are only required in the cavity to satisfy BS 5839-1;

c. the cavity is used as a plenum and the recommendations about recirculating air distribution systems in BS 5588-9:1999 are followed;

d. the surface of the material/product used in the construction of the cavity which is exposed in the cavity is Class 0 (National class) or Class B-s3, d2 or better (European class) and the supports and fixings in the cavity are of non-combustible construction;

e. the flame spread rating of any pipe insulation system is Class 1 or Class C-s3, d2 or better (European class) (see appendix A);

f. any electrical wiring in the void is laid in metal trays, or in metal conduit; and

g. any other materials in the cavity are of limited combustibility or Class A2 or better (European class) (see appendix A).

Note: When a classification includes 's3, d2', this means that there is no limit set for smoke production and/or flaming droplets/particles.

6.4.5 Construction and fixings for cavity barriers

Every cavity barrier should be constructed to provide at least 30 minutes fire resistance. It may be formed by any construction provided for another purpose if it meets the provisions for cavity barriers (see appendix A, table A1, item 13).

Cavity barriers in a stud wall or partition, or provided around openings may be formed of:

a. steel at least 0.5mm thick;

b. timber at least 38mm thick;

c. polythene-sleeved mineral wool, or mineral wool slab, in either case under compression when installed in the cavity; or

d. calcium silicate, cement-based or gypsum-based boards at least 12mm thick.

Note: Cavity barriers provided around openings may be formed by the window or door frame if the frame is constructed of steel or timber of the minimum thickness in a) or b) above as appropriate.

A cavity barrier should, wherever possible, be tightly fitted to a rigid construction and mechanically fixed in position. Where this is not possible (for example, in the case of a junction with slates, tiles, corrugated sheeting or similar materials) the junction should be fire-stopped. Provisions for fire-stopping are set out in section 6.5.

Cavity barriers should also be fixed so that their performance is unlikely to be made ineffective by:

a. movement of the building due to subsidence, shrinkage or temperature change and movement of the external envelope due to wind;

b. collapse in a fire of any services penetrating them;

c. failure in a fire of their fixings (but see note below); and

d. failure in a fire of any material or construction which they abut. (For example, if a suspended ceiling is continued over the top of a fire-resisting wall or partition and direct connection is made between the ceiling and the cavity barrier above the line of the wall or partition, premature failure of the cavity barrier can occur when the ceiling collapses. However, this may not arise if the ceiling is designed to provide fire protection of 30 minutes or more.)

Note: Where cavity barriers are provided in roof spaces, the roof members to which they are fitted are not expected to have any fire resistance – for the purpose of supporting the cavity barrier(s).

Any openings in a cavity barrier should be limited to those for:

• doors which have at least 30 minutes fire resistance (see appendix C, table C1, item 8) and are fitted in accordance with the provisions of appendix C;

- the passage of pipes which meet the provisions in section 6.5;
- the passage of cables or conduits containing one or more cables;
- openings fitted with a suitably mounted automatic fire damper (see section 6.5.3.1); and
- ducts which (unless they are fire-resisting) are fitted with a suitably mounted automatic fire damper where they pass through the cavity barrier.

6.5 Protection of openings and fire-stopping

6.5.1 Introduction

Sections 6.3 and 6.4 make provisions for fire-separating elements and set out the circumstances in which there may be openings in them. This section deals with the protection of openings in such elements.

If a fire-separating element is to be effective, every joint or imperfection of fit, or opening to allow services to pass through the element, should be adequately protected by sealing or fire-stopping so that the fire resistance of the element is not impaired.

The measures in this section are intended to delay the passage of fire. They generally have the additional benefit of retarding smoke spread, but the test specified in appendix A for integrity does not directly stipulate criteria for the passage of smoke.

Detailed guidance on door openings and fire doors is given in appendix C.

6.5.2 Openings for pipes

Pipes which pass through a fire-separating element (unless the pipe is in a protected shaft), should meet the appropriate provisions in alternatives A, B or C below.

For property protection, pipes and services that penetrate through compartment floors should be additionally enclosed within fire-resisting ducts.

6.5.2.1 Alternative A: Proprietary seals (any pipe diameter)

Provide a proprietary sealing system which has been shown by test to maintain the fire resistance of the wall, floor or cavity barrier.

Where non-metal pipes penetrate the element these shall be fitted with a heat activated closure system where test evidence shows the performance is suitable for the end use. Such devices shall be fixed back to the structure so they will not be readily exposed to fire. Although guidance in support of regulations permits the use of low melting point pipes of up to 40mm diameter to penetrate elements without the use of special sealing devices, solely for life safety purposes, the risk of disproportionate fire and smoke damage to the property and contents must be considered if this option is adopted.

Table 11 Maximum nominal internal diameter of pipes passing through a compartment wall/floor (see section 6.5.2 onwards)

Situation	Pipe material and maximum nominal internal diameter (mm)		
	(a) Non-combustible material[1]	(b) Lead, aluminium, aluminium alloy, uPVC[2], fibre cement	(c) Any other material
1. Structure (but not a wall separating buildings) enclosing a protected shaft which is not a stairway or a lift shaft	160	110	40
2. Any other situation	160	40	40

Notes:

1. Any non-combustible material (such as cast iron, copper or steel) which, if exposed to a temperature of 800°C, will not soften or fracture to the extent that flame or hot gas will pass through the wall of the pipe.

2. uPVC pipes complying with BS 4514:2001 and uPVC pipes complying with BS 5255:1989.

6.5.2.2 Alternative B: Pipes with a restricted diameter

Where a proprietary sealing system is not used, fire-stopping may be used around the pipe, keeping the opening as small as possible. The nominal internal diameter of the pipe should not be more than the relevant dimension given in table 11.

The diameters given in table 11 for pipes of specification (b) used in situation (2) assumes that the pipes are part of an above ground drainage system and are enclosed as shown in figure 35 if they are not, the smaller diameter given for situation (3) should be used.

6.5.2.3 Alternative C: Sleeving

A pipe of lead, aluminium, aluminium alloy, fibre-cement or uPVC, with a maximum nominal internal diameter of 160mm, may be used with a sleeving of non-combustible pipe as shown in figure 34. The specification for non-combustible and uPVC pipes is given in the notes to table 11.

Figure 34 Pipes penetrating structure

See section 6.5.2.3

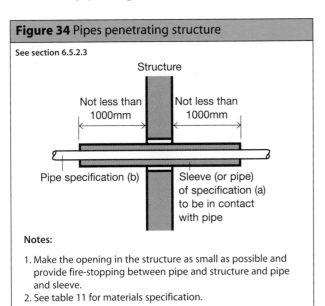

Structure

Not less than 1000mm | Not less than 1000mm

Pipe specification (b) | Sleeve (or pipe) of specification (a) to be in contact with pipe

Notes:

1. Make the opening in the structure as small as possible and provide fire-stopping between pipe and structure and pipe and sleeve.
2. See table 11 for materials specification.

6.5.3 Ventilation ducts, flues, etc.

Where air handling ducts pass through fire separating elements the integrity of those elements should be maintained.

There are three basic methods and these are:

Method 1 Protection using fire dampers;

Method 2 Protection using fire-resisting enclosures;

Method 3 Protection using fire-resisting ductwork.

Method 1 is not suitable for extract ductwork serving kitchens. This is due to the likely build up of grease within the duct which can adversely affect the effectiveness of any dampers.

Further information on fire-resisting ductwork is given in the ASFP Blue Book: *Fire-resisting ductwork* (ref).

6.5.3.1 Fire dampers

Fire dampers should be situated within the thickness of the fire-separating elements and be securely fixed. It is also necessary to ensure that, in a fire, expansion of the ductwork would not push the fire damper through the structure.

Adequate means of access should be provided to allow inspection, testing and maintenance of

Figure 35 Enclosure for drainage or water supply pipes

See section 6.5.2.2

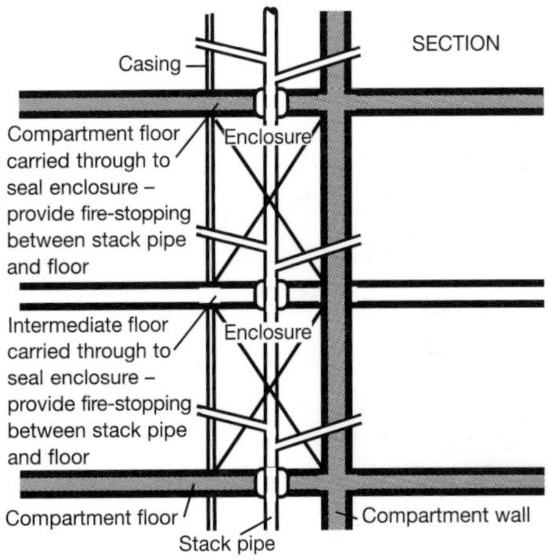

SECTION

Casing

Compartment floor carried through to seal enclosure – provide fire-stopping between stack pipe and floor | Enclosure

Intermediate floor carried through to seal enclosure – provide fire-stopping between stack pipe and floor | Enclosure

Compartment floor | Compartment wall

Stack pipe

Notes:

1. The enclosure should:
   a. be bounded by a compartment wall or floor, an outside wall, an intermediate floor or a casing (see specification in note 2);
   b. have internal surfaces (except framing members) of Class 0 (National class) or Class B-s3, d2 or better (European class). Note: When a classification includes 's3, d2', this means that there is no limit set for smoke production and/or flaming droplets/particles;
   c. not have an access panel which opens into a circulation space; and
   d. be used only for drainage, or water supply, or vent pipes for a drainage system.

2. The casing should:
   a. be imperforate except for an opening for a pipe or an access panel;
   b. not be of sheet metal; and
   c. have (including any access panel) not less than 30 minutes fire resistance.

3. The opening for a pipe, either in the structure or the casing, should be as small as possible and fire-stopped around the pipe.

both the fire damper and its actuating mechanism.

Further guidance on the design and installation of mechanical ventilation and air-conditioning plant is given in BS 5720:1979 on ventilation and on air-conditioning ductwork in BS 5588-9:1999.

Further information on fire and smoke-resisting dampers is given in the ASFP Grey Book: *Fire and smoke resisting dampers* (ref).

Fire dampers should be tested to BS EN 1366-2:1999 and be classified to BS EN 13501-3:2005. They should have an E classification equal to, or greater than, 60 minutes. Fire and smoke dampers should also be tested to BS EN 1366-2:1999 and be classified to BS EN 13501-3. They should have an ES classification equal to, or greater than, 60 minutes.

For property protection, fire dampers should also satisfy LPS 1162 (ref).

Note 1: Fire dampers tested using ad-hoc procedures based on BS 476 may only be appropriate for situations where the fan is switched off. In all cases, fire dampers should be installed as tested.

Note 2: Sections 4.5.7 and 6.3.5.4 also deal with ventilation and air-conditioning ducts.

6.5.3.2 Flues, etc.

If a flue, or duct containing flues or appliance ventilation duct(s), passes through a compartment wall or compartment floor, or is built into a compartment wall, each wall of the flue or duct should have a fire resistance of at least half that of the wall or floor in order to prevent the by-passing of the compartmentation (see figure 36).

For property protection, ducts must either have penetration seals, or else a fire resistance equal to that of the compartment wall or floor. Pipes carrying volatile liquids must be in a duct with 60 minutes fire resistance (integrity and insulation) and have penetration seals.

6.5.4 Fire-stopping

For fire separation to be effective, there should be continuity at the junctions of the fire-resisting elements enclosing a compartment or protected space, and any opening from one fire zone to another should not present a weakness.

Recommendations for fire stopping 'accidental' gaps between various building materials is summarised in table 12 and provides useful guidance on sealant materials. This table defines the use of rigid intumescent sealants (RI), flexible non-intumescent (FN) sealants and flexible intumescent (FI) sealants.

The recommendations made in the table take into account the response to fire of the materials bounding the gap, eg, whether they erode, shrink, expand, bow and the influence this will have on the seals.

However it must be pointed out that the orientation of the gap will also affect how well

---

Figure 36 Flues penetrating compartmental walls or floors (note that there is guidance in Approved Document J concerning hearths adjacent to compartment walls)

See section 6.5.3.2

a. Flue passing through compartment wall or floor

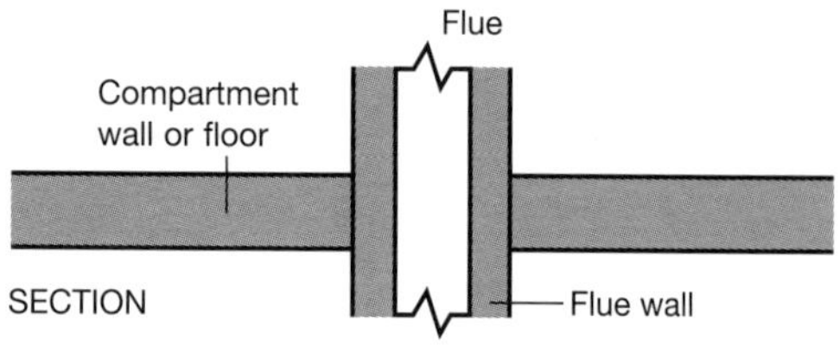

Flue walls should have a fire resistance of at least one half of that required for the compartment wall or floor, and be of non-combustible construction.

b. Flue built into compartment wall

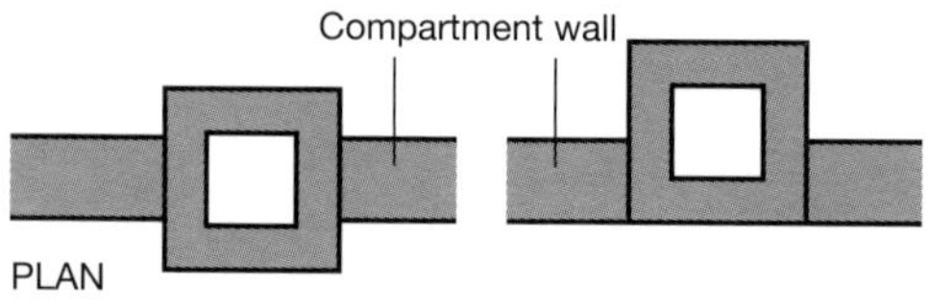

In each case flue wall should have a fire resistance at least one half of that required for the compartment wall or floor, and be of non-combustible construction.

the material stays in place and so the choice of product will also need to consider whether it has high or low adhesive qualities and whether the life of the seal can be influenced by the orientation.

The linear gap sealing material should be able to demonstrate by test evidence from BS EN 1366 Part 4, or assessment that it can maintain the integrity and insulation characteristics of the wall or the floor, or both, at the appropriate gap width, orientation and in the associated construction to be useful in practice. See guidance produced by Intumescent Fire Seals Association (IFSA) and ASFP (ref).

In addition to any other provisions in this document for fire-stopping:

a. joints between fire-separating elements should be fire-stopped; and

b. all openings for pipes, ducts, conduits or cables to pass through any part of a fire-separating element should be:

   i. kept as few in number as possible; and

   ii. kept as small as practicable; and

   iii. fire-stopped (which in the case of a pipe or duct, should allow thermal movement).

To prevent displacement, materials used for fire-stopping should be reinforced with (or supported by) materials of limited combustibility in the following circumstances:

• in all cases where the unsupported span is greater than 100mm; and

• in any other case where non-rigid materials are used (unless they have been shown to be satisfactory by test).

Proprietary fire-stopping and sealing systems (including those designed for service penetrations) which have been shown by test to maintain the fire resistance of the wall or other element, are available and may be used.

Other fire-stopping materials include:

• cement mortar;

• gypsum-based plaster;

• cement-based or gypsum-based vermiculite/perlite mixes;

• glass fibre, crushed rock, blast furnace slag or ceramic-based products (with or without resin binders); and

• intumescent mastics.

These may be used in situations appropriate to the particular material. Not all of them will be suitable in every situation.

Guidance on the process of design, installation and maintenance of passive fire protection is available in *Ensuring best practice for passive fire protection in buildings* (ref).

Further information on the generic types of systems available, information about their suitability for different applications and guidance on test methods is given in the ASFP Red Book: *Fire Stopping and Penetration Seals for the Construction Industry – the 'Red Book'* (ref).

Table 12 Recommended product selection for fire-stopping gaps (IFSA)						
	Masonry	Concrete	Timber	Gypsum	Fire protected steel	Steel
Masonry	RI/FN(2)	RI/FN(2)	FI	RI	RI(2)	*
Concrete		RI/FN(2)	FI	RI	RI(2)	*
Timber			RI	FI	FI(2)	*
Gypsum				FI	FI(2)	*
Fire Protected Steel(3)					RI(2)	FI(2)
Steel(1)	*	*	*	*	*	FN

(1) restricted to 30min applications

(2) whilst an RI material or FN material can be used, FI would be beneficial

(3) may be non-intumescent if protection does not degrade at all during heating

* treat as a functional linear gap seal

Section 7
External fire spread

7.1 Overview

7.1.1 Requirement B4 of the Building Regulations

(1) The external walls of the building shall adequately resist the spread of fire over the walls and from one building to another, having regard to the height, use and position of the building.

(2) The roof of the building shall adequately resist the spread of fire over the roof and from one building to another, having regard to the use and position of the building.

7.1.2 Performance

The Requirements of B4 will be met:

a. if the external walls are constructed so that the risk of ignition from an external source and the spread of fire over their surfaces, is restricted, by making provision for them to have low rates of heat release;

b. if the amount of unprotected area in the side of the building is restricted so as to limit the amount of thermal radiation that can pass through the wall, taking the distance between the wall and the boundary into account; and

c. if the roof is constructed so that the risk of spread of flame and/or fire penetration from an external fire source is restricted.

In each case so as to limit the risk of a fire spreading from the building to a building beyond the boundary, or vice versa.

The extent to which this is necessary is dependent on the use of the building, its distance from the boundary and, in some cases, its height, and the provision of fire suppression systems.

However, designers need to take account of the possibility that an external fire which penetrates the building may cause many sprinkler heads to activate, which could rapidly exhaust the water supply (sprinkler systems are normally designed to control fires originating from within buildings, when the activation of between 1~4 heads is usually sufficient to do the job).

7.1.3 Introduction

7.1.3.1 External walls

The construction of external walls and the separation between buildings to prevent external fire spread are closely related.

The chances of fire spreading across an open space between buildings and the consequences if it does, depend on:

- the size and intensity of the fire in the building concerned;
- the distance between the buildings;
- the fire protection given by their facing sides; and
- the risk presented to people in the other building(s).

The risk of fire transfer in the same building by the recognised mechanism of break out followed by break in via adjacent or upper windows should also be considered. Fire movement in this way can be minimised by the use of fire-resisting glazing systems in the external wall.

Provisions are made in section 7.2 for the fire resistance of external walls and to limit the susceptibility of the external surface of walls to ignition and to fire spread.

Provisions are made in section 7.3 to limit the extent of openings and other unprotected areas in external walls in order to reduce the risk of fire spread by radiation. This need not be the case, however, if insulating fire-resisting glass is used. For example, 30 minute integrity glass types are available which can be classified for 15 minutes insulation performance. The use of such glass types can widen the design options whilst maintaining natural lighting levels through maximum glazed areas.

7.1.3.2 Roofs

Provisions are made in section 7.4 for reducing the risk of fire spread between roofs and over the surfaces of roofs.

7.2 Construction of external walls

7.2.1 Introduction

Provisions are made in this section for the external walls of the building to have sufficient fire resistance to prevent fire spread across the relevant boundary. The provisions are closely linked with those for space separation in section 7.3 which sets out limits on the amount of unprotected area of wall. As the limits depend on the distance of the wall from the relevant boundary, it is possible for some or all of the walls to have no fire resistance, except for any parts which are load-bearing (see section 6.1.3.1).

External walls are elements of structure and the relevant period of fire resistance (specified in appendix A) depends on the use, height and size of the building concerned. If the wall is 1000mm or more from the relevant boundary, a reduced standard of fire resistance is accepted in most cases and the wall only needs fire resistance from the inside.

Provisions are also made to restrict the combustibility of external walls of schools. This is in order to reduce the surface's susceptibility to ignition from an external source and to reduce the danger from fire spread up the external face of the building.

In the guidance to Requirement B3, provisions are made in section 6.2 for internal and external load-bearing walls to maintain their load-bearing function in the event of fire.

7.2.2 Fire resistance standard

The external walls of the building should have the appropriate fire resistance given in appendix A, table A1, unless they form an unprotected area under the provisions of section 7.3.

7.2.3 External wall construction

The external envelope of a building should not provide a medium for fire spread if it is likely to be a risk to health or safety. The use of combustible materials in the cladding system and extensive cavities may present such a risk in tall buildings.

External walls should either meet the guidance given in section 7.2.4 and 7.2.4.1 or meet the performance criteria given in the BRE Report *Fire performance of external thermal insulation for walls of multi storey buildings* (BR 135) for cladding systems using full scale test data from BS 8414-1:2002 or BS 8414-2:2005.

The total amount of combustible material may also be limited in practice by the provisions for space separation in section 7.3 (see section 7.3.3 onwards).

7.2.4 External surfaces

The external surfaces of walls should meet the provisions in table 13.

To reduce the risk of external fires damaging the structure, combustible cladding for the ground floor level should only be used as part of a fire engineering solution.

Table 13 Provisions for external surfaces or walls		
Distance from relevant boundary	Height of wall	External wall surface classification
Less than 1000mm	Any	Class 0 (national standards) or class B-s3,d2 or better (European class). Profiled or flat steel sheet at least 0.5mm thick with an organic coating of no more than 0.2mm thickness is also acceptable
1000mm or more	Single storey	No requirement
1000mm or more	More than one storey: a) up to 10m above ground b) up to 10m above a roof or any part of the building to which pupils or the public have access	Index (I) not more than 20 (national class) or class C-s3,d2 or better (European class). Timber cladding at least 9mm thick is also acceptable. (The index I relates to tests specified in BS 476-6)

7.2.4.1 Cavity barriers

Cavity barriers should be provided in accordance with section 6.4.

In the case of an external wall construction, of a building which, by virtue of section 6.4.4.1 (external cladding system with a masonry or concrete inner leaf), is not subject to the provisions of table 10 (maximum dimensions of cavities), the surfaces which face into cavities should also meet the provisions of table 13.

7.3 Space separation

7.3.1 Introduction

The provisions in this section are based on a number of assumptions and, whilst some of these may differ from the circumstances of a particular case, together they enable a reasonable standard of space separation to be specified. The provisions limit the extent of unprotected areas in the sides of a building (such as openings and areas with a combustible surface) which will not give adequate protection against the external spread of fire from one building to another.

A roof is not subject to the provisions in this section unless it is pitched at an angle greater than 70° to the horizontal (see definition for 'external wall' in appendix F). Similarly, vertical parts of a pitched roof such as dormer windows (which taken in isolation might be regarded as a wall), would not need to meet the following provisions unless the slope of the roof exceeds 70°. It is a matter of judgement whether a continuous run of dormer windows occupying most of a steeply pitched roof should be treated as a wall rather than a roof.

The assumptions are:

a. that the size of a fire will depend on the compartmentation of the building, so that a fire may involve a complete compartment, but will not spread to other compartments;

b. that the intensity of the fire is related to the use of the building (ie, purpose group), but that it can be moderated by a sprinkler system;

c. that there is a building on the far side of the boundary that has a similar elevation to the one in question and that it is at the same distance from the common boundary; and

d. that the amount of radiation passing through any part of the external wall that has fire resistance be discounted provided that where glazing is concerned the fire-resisting glass has a measure of insulation performance for at least 15 minutes.

Where a reduced separation distance is desired (or an increased amount of unprotected area) it may be advantageous to construct compartments of a smaller size.

7.3.2 Boundaries

The use of the distance to a boundary, rather than to another building, in measuring the separation distance, makes it possible to calculate the allowable proportion of unprotected areas, regardless of whether there is a building on an adjoining site and regardless of the site of that building or the extent of any unprotected areas that it might have.

A wall is treated as facing a boundary if it makes an angle with it of 80° or less (see figure 37).

Usually only the distance to the actual boundary of the site needs to be considered. But in some circumstances, when the site boundary adjoins a space where further development is unlikely, such as a road, then part of the adjoining space may be included as falling within the relevant boundary for the purposes of this section. The meaning of the term boundary is explained in figure 37.

For property protection, the distance to notional boundaries (see section 7.3.2.2) between school buildings on the same site should be considered.

7.3.2.1 Relevant boundaries

The boundary which a wall faces, whether it is the actual boundary of the site or a notional boundary, is called the relevant boundary (see figures 37 and 38).

7.3.2.2 Notional boundaries

Notional boundaries are assumed to exist between different buildings on the same site.

The separation between buildings on the same site is normally only recommended for the purposes of property protection.

Figure 37 Relevant boundary

See sections 7.3.2.1 and 7.3.2.2

This diagram sets out the rules that apply in respect of a boundary for it to be considered as a relevant boundary.

For a boundary to be relevant it should:
a. coincide with; or
b. be parallel to; or
c. be at an angle of not more than 80° to the side of the building.

This boundary is at less than 80° to side C and is therefore relevant to side C

C

B Building D

A

This boundary is parallel to and therefore relevant to side D

This boundary coincides with and is therefore relevant to side B

This boundary is parallel to side A

But the relevant boundary may be the centre line of a road, railway, canal or river

However, if one of the buildings is residential, or where buildings on the site are operated/managed by different organisations, then separation is required to satisfy the Building Regulations with respect to life safety.

The appropriate rules are given in figure 38.

7.3.3   Unprotected areas and fire resistance

Any part of an external wall which has less fire resistance than the appropriate amount given in appendix A, table A2, is considered to be an unprotected area.

7.3.3.1   External walls of protected shafts forming stairways

Any part of an external wall of a stairway in a protected shaft is excluded from the assessment of unprotected area.

Note: There are provisions in the guidance to B1 (section 4.4.6.7, figure 22) which refers to section 2 of BS 5588-5:2004 about the relationship of external walls for protected stairways to the unprotected areas of other parts of the building.

Figure 38 Notional boundary

See section 7.3.2.2

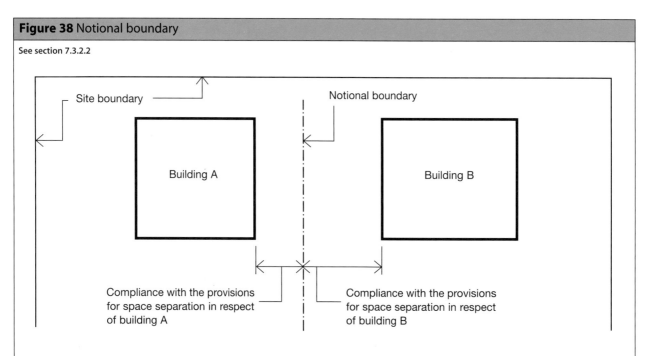

Site boundary

Notional boundary

Building A

Building B

Compliance with the provisions for space separation in respect of building A

Compliance with the provisions for space separation in respect of building B

The notional boundary should be set in the area between two buildings using the following rules:

1. The notional boundary is assumed to exist in the space between the buildings and is positioned so that one of the buildings would comply with the provisions of space separation having regard to the amount of its unprotected area. In practice, if one of the buildings is existing, the position of the boundary will be set by the space separation factors for that building.

2. The siting of the new building, or the second building if both are new, can then be checked to see that it also complies, using the notional boundary as the relevant boundary for the second building.

Building Regulations and Fire Safety Procedural Guidance, Fourth Edition

Communities and Local Government

This revised guide incorporates the Regulatory Reform Act and includes substantial changes to the previous version including new calculations on fire escape stairs, tighter requirements for fire-fighting shafts, more stringent demands for smoke control and allowances for sprinklers in lieu of alternative means of escape.

It has been prepared for designers, developers, occupiers, employers, fire and rescue authorities and building control bodies. It explains the steps involved in approving the fire safety aspects of building work, and the interaction between Building Regulations and other statutory fire safety requirements in England and Wales. If the procedures described in this guide are followed, developers and designers will be able to avoid abortive work. Owners and occupiers will also benefit, as this process will generate the basis of their fire safety management procedures and risk assessment.

Paperback
£8.50
40 pages
August 2007
ISBN 978 1 85946 2904
Code 62770

Guides to the Approved Documents

These three new easy-to-use, practical guides will help construction professionals to understand the complex provisions of Approved Documents B, L and F. Written in an uncluttered, user-friendly style, essential facts can be quickly assimilated through illustrations, flowcharts and watch-points. Key points and the mandatory requirements are highlighted to aid rapid digestion of the most important information. The authors of these guides have spent many man-hours on the reader's behalf looking into the meaning and importance of each clause, researching cross-references and compiling the guidance so that the reader doesn't have to.

Each guide will be will be an invaluable companion for construction professionals wanting reliable, straightforward guidance to the complex new Approved Documents.

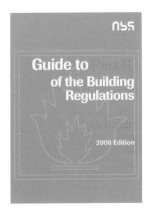

Guide to Part B
Simon Ham
Paperback
£25.00
176 pages
July 2007
ISBN 978 1 85946 271 3
Code 60834

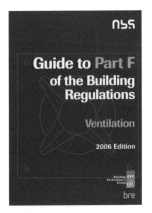

Guide to Part F
BPG (BRE)
Paperback
£17.50
56 pages
2006
ISBN 978 1 85946 198 3
Code 57271

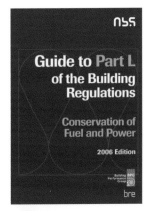

Guide to Part L
BPG (BRE)
Paperback
£27.50
152 pages
2006
ISBN 978 1 85946 199 0
Code 57270

Legislation Maze: Fire
Simon Ham

Paperback
£15.00
38 Pages
May 2007
ISBN 978 1 85946 251 5
Code 60749

This guide is a concise, highly accessible and up-to-date first point of reference for architects to all aspects of fire safety legislation. Organised according to the RIBA's Outline Plan of Work, the guidance covers both the 2006 edition of Part B of the Building Regulations and the significant new requirements of The Regulatory Reform (Fire Safety) Order 2005. Compliance with fire safety legislation can have a profound influence on the design of buildings and can result in considerable amount of abortive effort if an architect fails to address and incorporate solutions as the design develops. **Legislation Maze: Fire** is an essential desktop aide-memoire for those with responsibility for ensuring the life safety of their clients in the event of fire.

riba book shops. com

To order please copy the form, complete and return it to:

RIBA Bookshops Mail Order
15 Bonhill Street
London EC2P 2EA. UK
T +44 (0)20 7256 7222
F +44 (0)20 7374 2737
www.ribabookshops.com
Despatch
Mail order post and packing rates
UK
– Minimum postage charge is £2 for all orders under £20

– Under £100 add 10% to the total value of your order
– £100–£200 add £10 Flat Fee
– Over £200 items are sent post free
Overseas
– Minimum postage charge is £2.50 for all orders under £12.50
– Add 20% to the total value of your order

Online
– **UK orders over £50 are sent post free**
Please contact us for a quotation on premium delivery service. Please note that VAT is also charged on despatch on VATable items. VAT is charged at UK standard rate of 17.5%.

Order Form

Code	Title	Qty	Price	Total
63103	Building Bulletin 100: Design for fire safety in schools		£19.95	
62770	Building Regulations and Fire Safety Procedural Guidance		£8.50	
60834	Guide to Part B of the Building Regulations		£25.00	
57271	Guide to Part F of the Building Regulations		£17.50	
57270	Guide to Part L of the Building Regulations		£27.50	
60749	Legislation Maze: Fire		£15.00	
60509	Approved Document B: Volume 1 - Dwellinghouses		£12.50	
60511	Approved Document B: Volume 2 - Buildings other than dwellinghouses		£17.50	
		Sub-total		
		Despatch		
		Total		

Delivery address

Name

Address

Postcode

Telephone

Email

Date

Payment details

I enclose a cheque for £ _____ payable to RIBA Bookshops

Please debit my RIBA Bookshops account no _____

Please debit my ☐Mastercard ☐Visa ☐Switch ☐Maestro ☐Delta ☐Eurocard ☐Solo ☐American Express ☐Electron

Card number _____

Valid from _____ Expiry date _____ Issue no _____

Signature _____ Security code _____

Cardholder's address (if different from delivery address)

Data Protection
RIBA Enterprises would like to send you information about our products by post and email.
If you prefer not to receive this information please tick below:
☐ Please do not send information by email   ☐ Please do not send information by post

BB100A

Fire Safety
Design, Regulations & Guidance

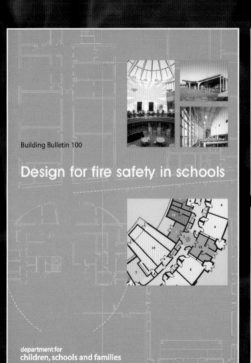

Building Bulletin 100

Design for fire safety in schools

department for
children, schools and families

∩bℲ the official publisher of the Approved
Documents to the Building Regulations

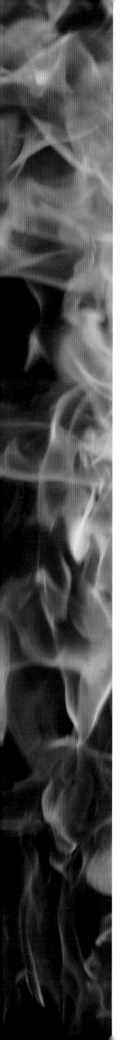

Building Bulletin 100:
Design for fire safety in schools

Department for Children, Schools and Families

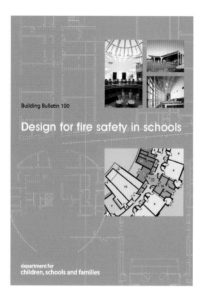

"What makes this different from previous fire safety guides is that it stresses the importance of protecting the fabric of schools. It shows how to protect school buildings from fire damage and includes extensive guidance on the use of sprinklers and their importance as a weapon against arson."

Jim Knight, MP
Minister of State for Schools and Learners
Department for Children, Schools and Families

Building Bulletin 100: Design for fire safety in schools is a landmark publication in improving fire safety in schools and clearly shows how the requirements for life safety, contained in the Building Regulations, can be met in the design of a new school or an extension. Innovative design may need to employ fire safety engineering, a risk-based approach, and this guide explains what design teams should do. It also covers the principles of fire safety management and describes the fire protection measures that the designer should consider.

The guide also acknowledges the important role of sprinklers. Sprinkler systems installed in buildings can significantly reduce the degree of damage caused by fire and can reduce the risk to life. On 1 March 2007, DCSF announced new policy on sprinklers and their value as a measure against the risk of fire and arson. The tools available to carry out such risk assessments are discussed in this guide. The first – which will enable you to determine whether an existing or proposed school is to be ranked high, medium or low risk, and the second – which is a cost benefit analysis assessment tool specifically covering the use of sprinklers – are both included in the CD-Rom which accompanies the guide.

The guide is broken down into 8 sections:

- Section 1 of the guide summaries the regulatory background and general guidance on fire safety in schools.

- Section 2 gives general design guidance and outlines the risk assessment to be used during the design process. This provides a way of selecting measures to control and manage those risks as a product of likelihood and impact of the risk identified.

- Sections 3-8 provide detailed design guidance that, if followed, will usually enable the school design to satisfy the requirements of B1-B5 of the Building Regulations.

The guide will be of use to all those with an interest in fire safety in schools, but in particular designers, fire engineers, building control officers and fire safety officers. Head teachers, governors, teaching staff and facilities and management staff will find it of interest to underpin their role as fire safety managers.

Paperback/160 pages/November 2007/**£19.95**/ISBN 978 1 85946 291 1/Code 63103

7.3.3.2 Status of combustible surface materials as unprotected area

If an external wall has the appropriate fire resistance, but has combustible material more than 1mm thick as its external surface, then that wall is counted as an unprotected area amounting to half the actual area of the combustible material, see figure 39. (For the purposes of this provision, a material with a Class 0 rating (National class) or Class B-s3, d2 rating (European class) (see appendix A) need not be counted as unprotected area).

Note: When a classification includes 's3, d2', this means that there is no limit set for smoke production and/or flaming droplets/particles.

7.3.3.3 Small unprotected areas

Small unprotected areas in an otherwise protected area of wall are considered to pose a negligible risk of fire spread and may be disregarded. Figure 40 shows the constraints that apply to the placing of such areas in relation to each other and to lines of compartmentation inside the building. These constraints vary according to the size of each unprotected area.

7.3.3.4 Canopies

Some canopy structures would be exempt from the application of the Building Regulations by falling within Class VI or Class VII of Schedule 2 to the Regulations (Exempt buildings and works). Many others may not meet the exemption criteria and in such cases the provisions in this section about limits of unprotected areas could be onerous.

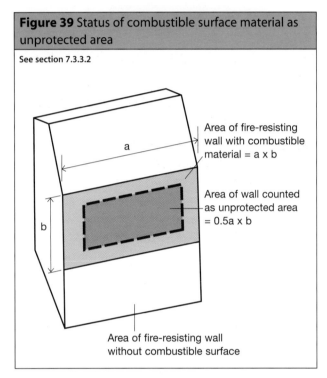

Figure 39 Status of combustible surface material as unprotected area

See section 7.3.3.2

Area of fire-resisting wall with combustible material = a x b

Area of wall counted as unprotected area = 0.5a x b

Area of fire-resisting wall without combustible surface

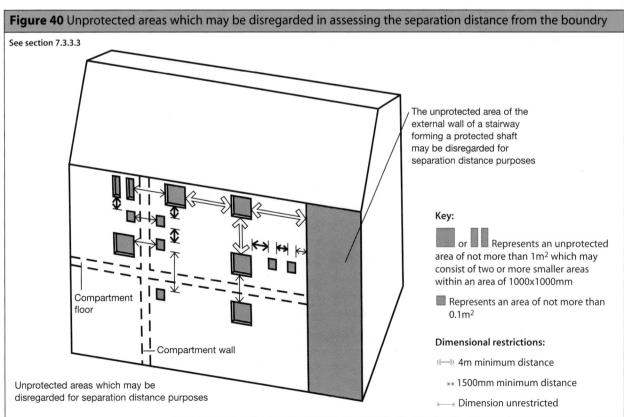

Figure 40 Unprotected areas which may be disregarded in assessing the separation distance from the boundry

See section 7.3.3.3

The unprotected area of the external wall of a stairway forming a protected shaft may be disregarded for separation distance purposes

Compartment floor

Compartment wall

Unprotected areas which may be disregarded for separation distance purposes

Key:

or ▪▪ Represents an unprotected area of not more than 1m² which may consist of two or more smaller areas within an area of 1000x1000mm

▪ Represents an area of not more than 0.1m²

Dimensional restrictions:

↤→ 4m minimum distance

↔ 1500mm minimum distance

↦——↤ Dimension unrestricted

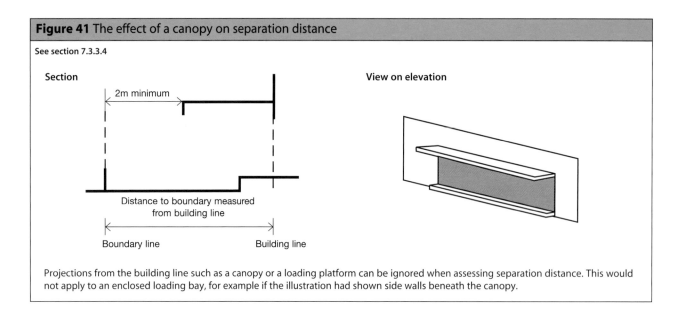

Figure 41 The effect of a canopy on separation distance

See section 7.3.3.4

Section

2m minimum

Distance to boundary measured from building line

Boundary line   Building line

View on elevation

Projections from the building line such as a canopy or a loading platform can be ignored when assessing separation distance. This would not apply to an enclosed loading bay, for example if the illustration had shown side walls beneath the canopy.

In the case of a canopy attached to the side of a building, provided that the edges of the canopy are at least 2m from the relevant boundary, separation distance may be determined from the wall rather than the edge of the canopy (see figure 41).

In the case of a free-standing canopy structure above a limited risk or controlled hazard, in view of the high degree of ventilation and heat dissipation achieved by the open sided construction and provided the canopy is 1000mm or more from the relevant boundary, the provisions for space separation could reasonably be disregarded.

7.3.3.5 External walls within 1000mm of the relevant boundary

A wall situated within 1000mm from any point on the relevant boundary and including a wall coincident with the boundary, will meet the provisions for space separation if:

a. the only unprotected areas are those shown in figure 40 (or part of the external wall of an uncompartmented building which are more than 30m above mean ground level, AD B, 13.12); and

b. the rest of the wall is fire-resisting from both sides.

7.3.3.6 External walls 1000mm or more from the relevant boundary

A wall situated at least 1000mm from any point on the relevant boundary will meet the provisions for space separation if:

- the extent of unprotected area does not exceed that given by one of the methods referred to in section 7.3.4; and
- the rest of the wall (if any) is fire-resisting from the inside of the building.

For buildings less than 10m apart, the fire resistance requirements are as given in table A2, appendix A (but note the higher fire resistance levels for property protection). These requirements also apply to walls at right angles to other buildings or walls overlooking a roof up to 10m below.

7.3.4 Methods for calculating acceptable unprotected area

A simple method is given in this Approved Document for calculating the acceptable amount of unprotected area in an external wall that is at least 1000mm from any point on the relevant boundary. (For walls within 1000mm of the boundary see section 7.3.3.5.)

This method may be used for most buildings or compartments, and is set out in section 7.3.4.4.

There are other more precise methods, described in a BRE report *External fire spread: Building separation and boundary distances* (BR 187, BRE 1991), which may be used instead of this simple method. The 'Enclosing Rectangle' and 'Aggregate Notional Area' methods are included in the BRE report.

7.3.4.1 Basis for calculating acceptable unprotected area

The basis of the method is set out in Fire Research Technical Paper No 5, 1963. This has been reprinted as part of the BRE report referred to in section 7.3.4. The aim is to ensure that the building is separated from the boundary by at least half the distance at which the total thermal radiation intensity received from all unprotected areas in the wall would be 12.6 kW/m² (in still air), assuming the radiation intensity at each unprotected area is 84 kW/m².

7.3.4.2 Sprinkler systems

If a building is fitted throughout with a sprinkler system, it is reasonable to assume that the intensity and extent of a fire will be reduced. If the sprinkler system qualifies as a 'life safety' system, as defined in the SSLD document on sprinklers in schools, then the boundary distance may be half that for an otherwise similar, but unsprinklered, building. This is subject to there being a minimum distance of 1m. Alternatively, the amount of unprotected area may be doubled if the boundary distance is maintained.

Note: The presence of sprinklers may be taken into account in a similar way when using the BRE report referred to in section 7.3.4.

7.3.4.3 Atrium buildings

If a building contains one or more atria, the recommendations of clause 28.2 in BS 5588-7:1997 should be followed.

7.3.4.4 Simple method

This method applies to a building or compartment intended for any use and which is not less than 1000mm from any point on the relevant boundary.

The following rules for determining the maximum unprotected area should be read with table 14.

The building or compartment should not exceed 10m in height.

Table 14 Permitted unprotected areas in small buildings or compartments

Minimum distance between side of building and relevant boundary (m)	Maximum total percentage of unprotected area (%)
1	8
2.5	20
5	40
7.5	60
10	80
12.5	100

Notes:

a. Intermediate values may be obtained by interpolation.

b. For buildings which are fitted throughout with an automatic sprinkler system, see section 7.3.4.2.

c. The total percentage of unprotected area is found by dividing the total unprotected area by the area of a rectangle that encloses all the unprotected areas and multiplying the result by 100.

Note: For any building or compartment more than 10m in height, the methods set out in the BRE report *External fire spread: Building separation and boundary distances can be applied.*

Each side of the building will meet the provisions for space separation if either:

i.   the distance of the side of the building from the relevant boundary; and

ii. the extent of unprotected area, are within the appropriate limits given in table 14.

Note: In calculating the maximum unprotected area, any areas shown in figure 40 and referred to in section 7.3.3.3, can be disregarded.

iii. any parts of the side of the building in excess of the maximum unprotected area should be fire-resisting.

Note: For the purposes of property protection, facilities for storage of combustible materials should not be located within 10m of the outside of the building

7.4 Roof coverings

7.4.1 Introduction

If a fire starts outside a building or breaks out of a building there is a danger that it might spread to adjacent buildings. For this reason regulatory guidance limits the use of roof coverings which will not give adequate protection against the spread of fire over them near a boundary (see table A5 and appendix A).

The term roof covering is used to describe constructions which may consist of one or more layers of material, but does not refer to the roof structure as a whole. The provisions in this section are principally concerned with the performance of roofs when exposed to fire from the outside.

The circumstances when a roof is subject to the provisions in section 7.3 for space separation are explained in section 7.3.1.

7.4.1.1   Other controls on roofs

There are provisions concerning the fire properties of roofs in three other sections of this document. In the guidance to B1 (section 4.5.2.1) there are provisions for roofs that are part of a means of escape. In the guidance to B2 (section 5.2.3.3) there are provisions for the internal surfaces of rooflights as part of the internal lining of a room or circulation space. In the guidance to B3 there are provisions in section 6.2.2.2 for roofs which are used as a floor and in section 6.3.3.6 for roofs that pass over the top of a compartment wall.

7.4.2   Classification of performance

The performance of roof coverings is designated by reference to the test methods specified in BS 476-3: 2004 or determined in accordance with BS EN 13501-5:2005, as described in appendix A. The notional performance of some common roof coverings is given in table A5 of appendix A.

Rooflights are controlled on a similar basis and plastic rooflights described in section 7.4.2.2 may also be used.

7.4.2.1   Separation distances

The separation distance is the minimum distance from the roof (or part of the roof) to the relevant boundary, which may be a notional boundary.

Table 15 sets out separation distances according to the type of roof covering and the size and use of the building. There are no restrictions on the use of roof coverings designated AA, AB or AC (National class) or $B_{ROOF}(t4)$ (European class) classification. In addition, roof covering products (and/or materials) as defined in Commission Decision 2000/553/EC of 6th September 2000 implementing Council Directive 89/106/EEC as

Table 15 Limitations on roof coverings*					
Designation† of covering of roof or part of roof		Minimum distance from any point on relevant boundary			
National Class	European Class	Less than 6m	At least 6m	At least 12m	At least 20m
AA, AB or AC	$B_{ROOF}(t4)$	✓	✓	✓	✓
BA, BB or BC	$C_{ROOF}(t4)$	✗	✓	✓	✓
CA, CB or CC	$D_{ROOF}(t4)$	✗	✓ (1)(2)	✓ (1)	✓
AD, BD or CD	$E_{ROOF}(t4)$	✗	✓	✓ (1)	✓
DA, DB, DC or DD	$F_{ROOF}(t4)$	✗	✗	✗	✓ (1)(2)

Notes:

*   See section 7.4.2.3 for limitations on glass, section 7.4.2.4 for limitations on thatch and wood shingles; and section 7.4.2.2 and tables 16 and 17 for limitations on plastic rooflights.

†   The designation of external roof surfaces is explained in appendix A. (See table A5, for notional designations of roof coverings.)

Openable polycarbonate and PVC rooflights which achieve a Class 1 (National class) or Class C-s3, d2 (European class) rating by test, see section 7.4.2.2, may be regarded as having an AA (National class) designation or $B_{ROOF}(t4)$ (European class) classification.

✔   Acceptable.
✗   Not acceptable.

1. Not acceptable on any buildings with a cubic capacity of more than 1500m³.

2. Acceptable on buildings not listed in Note 1, if part of the roof is no more than 3m² in area and is at least 1500mm from any similar part, with the roof between the parts covered with a material of limited combustibility.

regards the external fire performance of roof coverings can be considered to fulfil all of the requirements for performance characteristic 'external fire performance' without the need for testing provided that any national provisions on the design and execution of works are fulfilled. That is, the roof covering products (and/or materials) defined in this Commission Decision can be used without restriction.

Where it is desired to protect high value/business-critical areas, each rooflight should have a maximum dimension of 1m, rather than an area of not more than 5m².

7.4.2.2   Plastic rooflights

Table 16 sets out the limitations on the use of plastic rooflights which have at least a Class 3 (National class) or Class D-s3, d2 (European class) lower surface and table 17 sets out the limitations on the use of thermoplastic materials with a TP(a) rigid or TP(b) classification (see also figure 42). The method of classifying thermoplastic materials is given in appendix A.

Figure 42 Limitations on spacing and size of plastic rooflights having a Class 3 (National class) or Class D-s3, d2 (European class) or TP(b)lower surface

See section 7.4.2.2

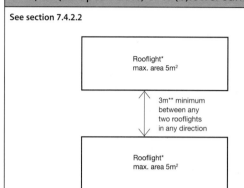

* Or group of rooflights amounting to no more than 5m².

** Class 3 rooflights may be spaced 1800mm apart provided the rooflights are evenly distributed and do not exceed 20% of the area of the room.

Notes:

1. There are restrictions on the use of plastic rooflights in the guidance of section 5.

2. Surrounding roof covering to be a material of limited combustibility for at least 3m distance.

3. Where figure 28a or b applies, rooflights should be at least 1500mm from the compartment wall.

Table 16 Class 3 (National class) or Class D-s3, d2 (European class) plastic rooflights: limitations on use and boundary distance			
Minimum classification on lower surface[1]	Space which rooflight can serve	Minimum distance from any point on relevant boundary to rooflight with an external designation[†] of:	
		AD BD CD (National class) or E_{ROOF}(t4) (European class) CA CB CC or D_{ROOF}(t4) (European class)	DA DB DC DD (National class) or F_{ROOF}(t4) (European class)
Class 3	a. Balcony, verandah, carport, covered way or loading bay, which has at least one longer side wholly or permanently open	6m	20m
	b. Detached swimming pool		
	c. Conservatory, garage or outbuilding, with a maximum floor area of 40m²		
	d. Circulation space[2] (except a protected stairway)	6m[3]	20m[3]
	e. Room[2]		

Notes:

† The designation of external roof surfaces is explained in appendix A.

None of the above designations are suitable for protected stairways – see section 5.2.3.3.

Polycarbonate and PVC rooflights which achieve a Class 1 (National class) or Class C-s3, d2 (European class) rating by test, see section 7.4.2.2, may be regarded as having an AA designation or B_{ROOF}(t4) (European class) classification.

Where figure 28a or b applies, rooflights should be at least 1.5m from the compartment wall.

Products may have upper and lower surfaces with different properties if they have double skins or are laminates of different materials. In which case the more onerous distance applies.

1. See also the guidance to B2 (see section 5.2.2.4 and 5.2.3.3).

2. Single skin rooflight only, in the case of non-thermoplastic material.

3. The rooflight should also meet the provisions of figure 42.

Aside: it should not be assumed that collapsed plastic rooflights will provide adequate ventilation in the event of a fire. A properly-designed smoke extract system should be used instead, if required.

When used in rooflights, a rigid thermoplastic sheet product made from polycarbonate or from unplasticised PVC, which achieves a Class 1 (National class) rating for surface spread of flame when tested to BS 476-7:1971 (or 1987 or 1997) *Surface spread of flame tests for materials*, or Class C-s3, d2 (European class) can be regarded as having an AA (National class) designation or B_{ROOF}(t4) (European class) classification, other than for the purposes of figure 28.

7.4.2.3 Unwired glass in rooflights

When used in rooflights, unwired glass at least 4mm thick can be regarded as having an AA designation (National class) designation or B_{ROOF}(t4) (European class) classification.

7.4.2.4 Thatch and wood shingles

Thatch and wood shingles should be regarded as having an AD/BD/CD designation or E_{ROOF}(t4) (European class) classification in table 15 if performance under BS 476-3:1958 2004 or EN 1187:XXX (test 4) respectively cannot be established.

Table 17 TP(a) and TP(b) plastic rooflights: limitations on use and boundary distance			
Minimum classification on lower surface[1]	Space which rooflight can serve	Minimum distance from any point on relevant boundary to rooflight with an external designation[1] of:	
		TP(a)	TP(b)
1. TP(a) rigid	Any space except a protected stairway	6m[2]	Not applicable
2. TP(b)	a. Balcony, verandah, carport, covered way or loading bay, which has at least one longer side wholly or permanently open b. Detached swimming pool c. Conservatory, garage or outbuilding, with a maximum floor area of 40m²	Not applicable	6m
	d. Circulation space[3] (except a protected stairway) e. Room[3]	Not applicable	6m[4]

Notes:

None of the above designations are suitable for protected stairways – see section 5.2.3.3.

Polycarbonate and PVC rooflights which achieve a Class 1 rating by test, see section 7.4.2.2, may be regarded as having an AA designation.

Where figure 28a or b applies, rooflights should be at least 1.5m from the compartment wall.

Products may have upper and lower surfaces with different properties if they have double skins or are laminates of different materials; in which case the more onerous distance applies.

1. See also the guidance to B2 (see section 5.2.2.4 and 5.2.3.3).

2. No limit in the case of any space described in 2a, b and c.

3. Single skin rooflight only, in the case of non-thermoplastic material.

4. The rooflight should also meet the provisions of figure 42.

Section 8
Access and facilities for the Fire and Rescue Service

8.1 Overview

8.1.1 Requirement B5 of the Building Regulations

(1) The building shall be designed and constructed so as to provide reasonable facilities to assist firefighters in the protection of life.

(2) Reasonable provision shall be made within the site of the building to enable fire appliances to gain access to the building.

8.1.2 Performance

The Requirements of B5 will be met:

a. if there is sufficient means of external access to enable fire appliances to be brought near to the building for effective use;

b. if there is sufficient means of access into and within, the building for fire-fighting personnel to effect search and rescue and fight fire;

c. if the building is provided with sufficient internal fire mains and other facilities to assist fire-fighters in their tasks; and

d. if the building is provided with adequate means for venting heat and smoke from a fire in a basement.

These access arrangements and facilities are only required in the interests of the health and safety of people in and around the building. The extent to which they are required will depend on the use and size of the building in so far as it affects the health and safety of those people.

8.1.3 Introduction

The guidance given here covers the selection and design of facilities for the purpose of protecting life by assisting the Fire and Rescue Service. To assist the Fire and Rescue Service some or all of the following facilities may be necessary, depending mainly on the size of the building:

• vehicle access for fire appliances;

• access for fire-fighting personnel;

• the provision of fire mains within the building;

• venting for heat and smoke from basement areas; and

• the provision of adequate water supplies.

If it is proposed to deviate from the general guidance in sections 8.2 – 8.5 then it would be advisable to seek advice from the relevant Fire and Rescue Service at the earliest opportunity, even where there is no statutory duty to consult.

8.1.3.1 Facilities appropriate to a specific building

The main factor determining the facilities needed to assist the Fire and Rescue Service is the size of the building. Generally speaking fire-fighting is carried out within the building.

They need special access facilities (see section 8.4), equipped with fire mains (see section 8.2). Fire appliances will need access to entry points near the fire mains (see section 8.3).

In most buildings, the combination of personnel access facilities offered by the normal means of escape and the ability to work from ladders and appliances on the perimeter, will generally be adequate without special internal arrangements. Vehicle access may be needed to some or all of the perimeter, depending on the size of the building (see section 8.2).

Note: Where an alternative approach outside the scope of this document has been used to justify the means of escape it may be necessary to consider additional provisions for fire-fighting access.

For small buildings, it is usually only necessary to ensure that the building is sufficiently close to a point accessible to Fire and Rescue Service vehicles (see section 8.3.2).

Products of combustion from basement fires tend to escape via stairways, making access difficult for Fire and Rescue Service personnel. The problem can be reduced by providing vents (see section 8.4). Venting can improve visibility and reduce temperatures, making search, rescue and fire-fighting less difficult.

8.1.3.2 Insulating core panels

Guidance on the fire behaviour of insulating core panels used for internal structures is given in appendix B.

8.2   Fire mains and hydrants

8.2.1   Introduction

Fire mains are installed in a building and equipped with valves, etc, so that the Fire and Rescue Service may connect hoses for water to fight fires inside the building.

Fire mains may be of the 'dry' type which are normally empty and are supplied through a hose from a Fire and Rescue Service pumping appliance. Alternatively, they may be of the 'wet' type where they are kept full of water and supplied from tanks and pumps in the building. There should be a facility to allow a wet system to be replenished from a pumping appliance in an emergency.

8.2.2   Fire mains in buildings with fire-fighting shafts

Buildings with fire-fighting shafts should be provided with fire mains in those shafts and, where necessary, in protected escape stairs. The criteria for the provision of fire-fighting shafts and fire mains in such buildings are given in section 8.4.

8.2.2.1   Fire mains in other buildings

Fire mains may also be provided in other buildings where vehicle access is not provided in accordance with table 18 (see section 8.3.3) or section 8.3.2.

8.2.3   Number and location of fire mains

In buildings provided with fire mains for the purposes of section 8.2.2.1, outlets from fire mains should be located to meet the hose criterion set out in section 8.4.3. This does not imply that these stairs need to be designed as fire-fighting shafts.

8.2.4   Design and construction of fire mains

The outlets from fire mains should be located within the protected enclosure of a stairway or a protected lobby where one is provided (see figure 46).

Guidance on other aspects of the design and construction of fire mains, not included in the provisions of this document, should be obtained from BS 9990:2006.

8.2.5   Provision of private hydrants

Where a building, which has a compartment of 280m^2 or more in area, is being erected more than 100m from an existing fire-hydrant additional hydrants should be provided as follows;

*   **Buildings provided with fire mains** – hydrants should be provided within 90m of dry fire main inlets.

*   **Buildings not provided with fire mains** – hydrants should be provided within 90m of an entry point to the building and not more than 90m apart.

Each fire hydrant should be clearly indicated by a plate, affixed nearby in a conspicuous position, in accordance with BS 3251:1976.

Where no piped water supply is available, or there is insufficient pressure and flow in the water main, or an alternative arrangement is proposed, the alternative source of supply should be provided in accordance with the following recommendations:

a. a charged static water tank of at least 45,000 litre capacity; or

b. a spring, river, canal or pond capable of providing or storing at least 45,000 litres of water at all times of the year, to which access, space and a hard standing are available for a pumping appliance; or

c. any other means of providing a water supply for fire-fighting operations considered appropriate by the Fire and Rescue Authority.

8.3   Vehicle access

8.3.1   Introduction

For the purposes of this document vehicle access to the exterior of a building is needed to enable high reach appliances, such as turntable ladders and hydraulic platforms, to be used and to enable pumping appliances to supply water and equipment for fire-fighting, search and rescue activities.

Access requirements increase with building size and height.

Fire mains (see section 8.2) enable fire-fighters within the building to connect their hoses to a water supply. In buildings fitted with fire mains, pumping appliances need access to the perimeter at points near the mains, where fire-fighters can enter the building and where in the case of dry mains, a hose connection will be made from the appliance to pump water into the main.

The vehicle access requirements described in table 18 for buildings without fire mains, do not apply to buildings with fire mains.

Vehicle access routes and hard-standings should meet the criteria described in sections 8.3.3 – 8.3.4 where they are to be used by Fire and Rescue Service vehicles.

Note: Requirements cannot be made under the Building Regulations for work to be done outside the site of the works shown on the deposited plans, building notice or initial notice. In this connection it may not always be reasonable to upgrade an existing route across a site to a small building. The options in such a case, from doing no work to upgrading certain features of the route, eg, a sharp bend, should be considered by the designers in consultation with the Building Control Body and the Fire and Rescue Service.

8.3.2 Buildings not fitted with fire mains

8.3.2.1 Small buildings

There should be vehicle access for a pump appliance to small buildings (those of up to 2,000m² with a top storey up to 11m above ground level) to either:

- 15% of the perimeter; or
- within 45m of every point on the projected plan area (or 'footprint', see figure 43) of the building; whichever is the less onerous.

Note 1: If these provisions cannot be met, a fire main should be provided in accordance with section 8.2.2.1 and vehicle access should meet section 8.3.3.

8.3.2.2 Large buildings

Vehicle access to buildings that do not have fire mains (other than small buildings described in section 8.3.2.1) should be provided in accordance with table 18.

Every elevation to which vehicle access is provided in accordance with this section or

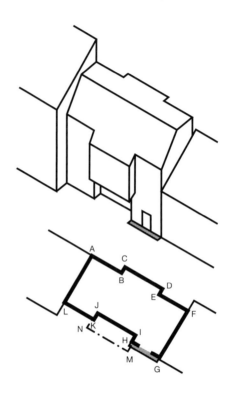

Figure 43 Example of building footprint and perimeter

See section 8.3.2 and table 18

Plan of building AFGL, where AL and FG are walls in common with other buildings.

The footprint of the building is the maximum aggregate plan perimeter found by the vertical projection of any overhanging storey onto a ground storey (ie, ABCDEFGHMNKL).

The perimeter of the footprint for the purposes of table 18 is the sum of the lengths of the two external walls, taking account of the footprint, ie, (A to B to C to D to E to F) + (G to H to M to N to K to L).

If the dimensions of the building are such that table 18 requires vehicle access, the shaded area illustrates one possible approach to 15% of the perimeter. Note: there should be a door into the building in this length (see section 8.3.2.1).

If the building does not have walls in common with other buildings, the lengths AL and FG would be included in the perimeter.

table 18 should have a suitable door(s), not less than 750mm wide, giving access to the interior of the building.

8.3.2.3 Access doors for fire-fighters

Door(s) should be provided such that there is no more than 60m between each door and/or the end of that elevation (eg, a 150m elevation would need at least 2 doors).

Table 18 Fire and Rescue Service vehicle access to school buildings not fitted with fire mains

Total floor area[1] of building m²	Height of floor of top storey above ground	Provide vehicle access[2][4] to:	Type of Appliance
up to 2,000	Up to 11	See Note[3]	Pump
	Over 11	15% of perimeter[5]	High reach
2,000-8,000	Up to 11	15% of perimeter[5]	Pump
	Over 11	50% of perimeter[5]	High reach
8,000-16,000	Up to 11	50% of perimeter[5]	Pump
	Over 11	50% of perimeter[5]	High reach
16,000-24,000	Up to 11	75% of perimeter[5]	Pump
	Over 11	75% of perimeter[5]	High reach
over 24,000	Up to 11	100% of perimeter[5]	Pump
	Over 11	100% of perimeter[5]	High reach

Notes:

1. The total floor area is the aggregate of all floors in the building (excluding basements).
2. An access door not less than 750mm wide is required to each elevation to which vehicle access is provided, which gives access to the interior of the building.
3. Access to be provided for a pump appliance to either (a) 15% of the perimeter, or (b) within 45m of every point on the projected plan area of the building whichever is the less onerous.
4. See section 8.3.3 for meaning of access.
5. Perimeter is described in figure 43.

Table 19 Typical Fire and Rescue Service vehicle access route specification

Appliance type	Minimum width of road between kerbs (m)	Minimum width of gateways (m)	Minimum turning circle between kerbs (m)	Minimum turning circle between walls (m)	Minimum carrying capacity (tonnes)
Pump	3.7	3.1	16.8	19.2	12.5
High Reach	3.7	3.1	26.0	29.0	17.0

Notes:

1. Fire appliances are not standardised. Some fire services have appliances of greater weight or different size. In consultation with the Fire Authority, the Designers and the Building Control Body may adopt other dimensions in such circumstances.
2. Because the weight of high reach appliances is distributed over a number of axles, and they are only used occasionally they should not cause any damage to a carriageway or route designed to 12.5 tonnes rather than the full 17 tonnes capacity.

Figure 44 Relationship between building and hardstanding/access roads for high reach fire appliances

See section 8.3.4

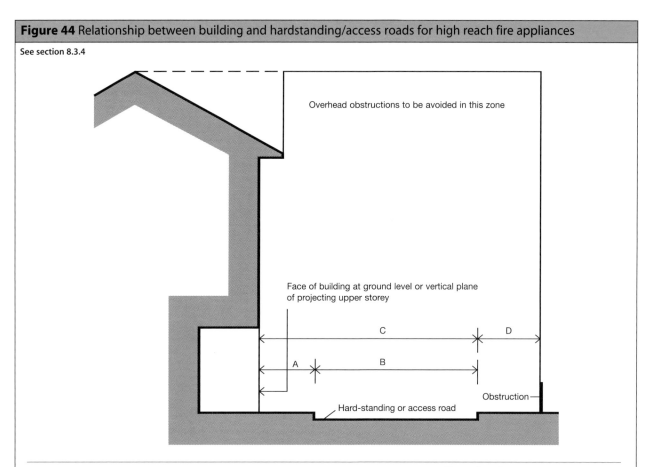

	Type of appliance	
	Turntable ladder dimension (m)	**Hydraulic platform dimension (m)**
A. Maximum distance of near edge of hardstanding from building	4.9	2.0
B. Minimum width of hardstanding	5.0	5.5
C. Minimum distance of further edge of hardstanding from building	10.0	7.5
D. Minimum width of unobstructed space (for swing of appliance platform)	N/A	2.2

Notes:

1. Hard-standing for high reach appliances should be as level as possible and should not exceed a gradient of 1 in 12.

2. Fire appliances are not standardised. Some fire services have appliances with a greater weight or different size. In consultation with the Fire and Rescue Service, the Building Control Body should adopt the relevant dimensions and ground loading capacity.

Figure 45 Turning facilities

See section 8.3.4

Fire and Rescue Service vehicles should not have to reverse more than 20m from the end of an access road.

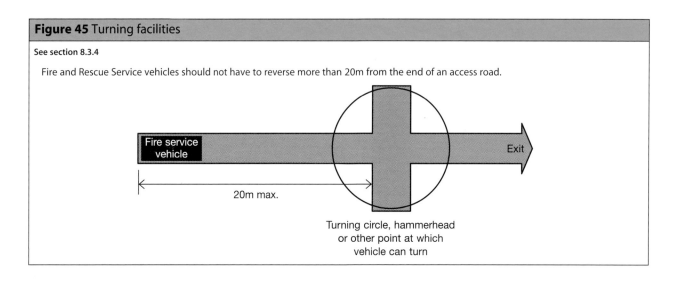

8.3.3   Buildings fitted with fire mains

In the case of a building fitted with dry fire mains there should be access for a pumping appliance to within 18m of each fire main inlet connection point, typically on the face of the building. The inlet should be visible from the appliance.

In the case of a building fitted with wet mains the pumping appliance access should be to within 18m and within sight of, a suitable entrance giving access to the main and in sight of the inlet for the emergency replenishment of the suction tank for the main.

Note: Where fire mains are provided in buildings for which sections 8.2 and 8.4 make no provision, vehicle access may be to this section rather than table 18.

8.3.4   Design of access routes and hard-standings

A vehicle access route may be a road or other route which, including any inspection covers and the like, meets the standards in table 19 and the following paragraphs.

Where access is provided to an elevation in accordance with table 18 for:

a. buildings up to 11m in height (excluding small buildings covered by section 8.3.2.1), there should be access for a pump appliance adjacent to the building for the percentage of the total perimeter specified;

b. buildings over 11m in height, the access routes should meet the guidance in figure 44.

Where access is provided to an elevation for high reach appliances in accordance with table 18, overhead obstructions such as cables and branches that would interfere with the setting of ladders, etc, should be avoided in the zone shown in figure 44.

Turning facilities should be provided in any dead-end access route that is more than 20m long (see figure 45). This can be by a hammerhead or turning circle, designed on the basis of table 19.

8.4   Access to buildings for fire-fighting personnel

8.4.1   Introduction

In low-rise buildings without deep basements Fire and Rescue Service personnel access requirements will be met by a combination of the normal means of escape and the measures for vehicle access in section 8.3, which facilitate ladder access to upper storeys. In other buildings, the problems of reaching the fire and working inside near the fire, necessitate the provision of additional facilities to avoid delay and to provide a sufficiently secure operating base to allow effective action to be taken.

These additional facilities include fire-fighting lifts, fire-fighting stairs and fire-fighting lobbies, which are combined in a protected shaft known as the fire-fighting shaft (figure 46).

Guidance on protected shafts in general is given in section 6.3.

Figure 46 Components of a fire-fighting shaft

See section 8.4.1

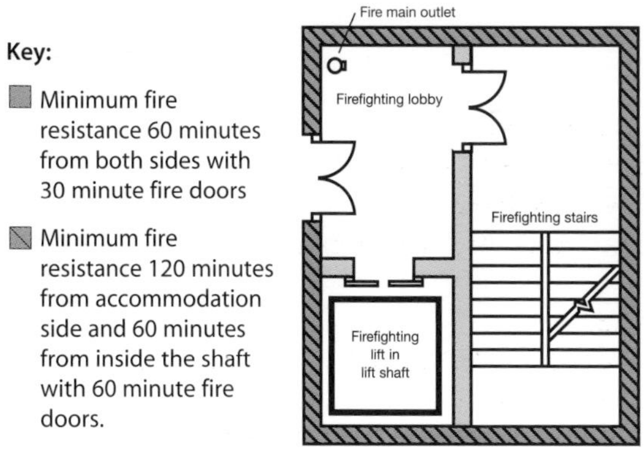

Key:

▨ Minimum fire resistance 60 minutes from both sides with 30 minute fire doors

▨ Minimum fire resistance 120 minutes from accommodation side and 60 minutes from inside the shaft with 60 minute fire doors.

Fire main outlet
Firefighting lobby
Firefighting stairs
Firefighting lift in lift shaft

Notes:

1. Outlets from a fire main should be located in the fire-fighting lobby.

2. Smoke control should be provided in accordance with BS 5588-5:2004.

3. A fire-fighting lift is required if the building has a floor more that 18m above or 10m below, fire service vehicle access level.

4. This figure is only to illustrate the basic components and is not meant to represent the only acceptable layout. The shaft should be constructed generally in accordance with clauses 7 and 8 of BS 5588-5:2004.

8.4.2   Provision of fire-fighting shafts

Fire-fighting shafts are provided to give fire-fighters a safe protected route from the outside of a building - and more importantly a safe exit route, if required – to (or from) all the upper or below-ground floors. Fire-fighting shafts will always contain a fire main which serves all floors above (dry rising main) and below ground (dry falling main). The dry rising main allows water to be quickly pumped to any floor without fire-fighters having to drag heavy hoses up stairs or haul them aloft from outside the building.

Buildings with a storey of 900m² or more in area, where the floor is at a height of more than 7.5m above Fire and Rescue Service vehicle access level, should be provided with fire-fighting shaft(s), which need not include fire-fighting lifts.

Buildings with two or more basement storeys, each exceeding 900m² in area, should be provided with fire-fighting shaft(s), which need not include fire-fighting lifts.

If a fire-fighting shaft is required to serve a basement it need not also serve the upper floors unless they also qualify because of the height or size of the building. Similarly a shaft serving upper storeys need not serve a basement which is not large or deep enough to qualify in its own right. However, a fire-fighting stair and any fire-fighting lift should serve all intermediate storeys between the highest and lowest storeys that they serve.

Fire-fighting shafts should serve all floors through which they pass.

8.4.3   Number and location of fire-fighting shafts

Fire-fighting shafts should be located to meet the maximum hose distances set out in this section and at least two should be provided in buildings with a storey of 900m² or more in area, where the floor is at a height of more than 7.5m above Fire and Rescue Service vehicle access level (see section 8.4.2).

If the building is fitted throughout with an automatic sprinkler system that qualifies as a 'life safety' system, as defined in the SSLD document on sprinklers in schools, then sufficient fire-fighting shafts should be provided such that every part of every storey, that is more

than 7.5m above Fire and Rescue Service vehicle access level (where covered by section 8.4.2), is no more than 60m from a fire main outlet in a fire-fighting shaft, measured on a route suitable for laying hose.

If the building is not fitted with sprinklers then every part of every storey that is more than 7.5m above Fire and Rescue Service vehicle access level (where covered by section 8.4.2), should be no more than 45m from a fire main outlet contained in a protected stairway and 60m from a fire main in a fire-fighting shaft, measured on a route suitable for laying hose.

Note: In order to meet the 45m hose criterion it may be necessary to provide additional fire mains in escape stairs. This does not imply that these stairs need to be designed as fire-fighting shafts.

8.4.4   Design and construction of fire-fighting shafts

Every fire-fighting stair and fire-fighting lift should be approached from the accommodation, through a fire-fighting lobby.

All fire-fighting shafts should be equipped with fire mains having outlet connections and valves at every storey.

A fire-fighting lift installation includes the lift car itself, the lift well and any lift machinery space, together with the lift control system and the lift communications system. The shaft should be constructed generally in accordance with clauses 7 and 8 of BS 5588-5:2004. Firefighting lift installations should conform to BS EN 81-72:2003 and to BS EN 81-1:1998 or BS EN 81 -2: 1998 as appropriate for the particular type of lift.

8.4.5   Rolling shutters in compartment walls

Rolling shutters should be capable of being opened and closed manually by the Fire and Rescue Service without the use of a ladder.

8.5   Venting of heat and smoke from basements

8.5.1   Introduction

The build-up of smoke and heat as a result of a fire can seriously inhibit the ability of the Fire and Rescue Service to carry out rescue and fire-

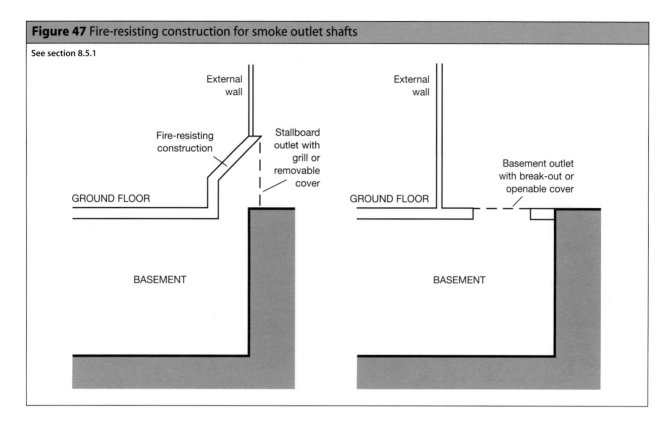

Figure 47 Fire-resisting construction for smoke outlet shafts

See section 8.5.1

fighting operations in a basement. The problem can be reduced by providing facilities to make conditions tenable for fire-fighters.

Smoke outlets (also referred to as smoke vents) provide a route for heat and smoke to escape to the open air from the basement level(s). They can also be used by the Fire and Rescue Service to let cooler air into the basement(s). (See figure 47).

8.5.2 Provision of smoke outlets

Where practicable each basement space should have one or more smoke outlets, but it is not always possible to do this where, for example, the plan is deep and the amount of external wall is restricted by adjoining buildings. It is therefore acceptable to vent spaces on the perimeter and allow other spaces to be vented indirectly by opening connecting doors. However if a basement is compartmented, each compartment should have direct access to venting, without having to open doors, etc, into another compartment.

Smoke outlets, connected directly to the open air, should be provided from every basement storey, except for any basement storey that has:

- a floor area of not more than 200m²; and
- a floor not more than 3m below the adjacent ground level.

Where basements have external doors or windows, the compartments containing the rooms with these doors or windows do not need smoke outlets. It is common for basements to be open to the air on one or more elevations. This may be the result of different ground levels on different sides of the building.

8.5.2.1 Natural smoke outlets

Smoke outlets should be sited at high level, either in the ceiling or in the wall of the space they serve. They should be evenly distributed around the perimeter to discharge in the open air outside the building.

The combined clear cross-sectional area of all smoke outlets should not be less than 1/40th of the floor area of the storey they serve.

Separate outlets should be provided from places of special fire hazard.

If the outlet terminates at a point that is not readily accessible, it should be kept unobstructed and should only be covered with a non-combustible grille or louvre.

If the outlet terminates in a readily accessible position, it may be covered by a panel, stallboard or pavement light which can be broken out or opened. The position of such covered outlets should be suitably indicated.

Outlets should not be placed where they would prevent the use of escape routes from the building.

8.5.2.2   Mechanical smoke extract

A system of mechanical extraction may be provided as an alternative to natural venting to remove smoke and heat from basements, provided that the basement storey(s) are fitted with a sprinkler system that qualifies as a 'life safety' system, as defined in the SSLD document on sprinklers in schools (it is not considered necessary in this particular case to install sprinklers on the storeys other than the basement(s) unless they are needed for other reasons).

The air extraction system should give at least 10 air changes per hour and should be capable of handling gas temperatures of 300°C for not less than one hour. It should come into operation automatically on activation of the sprinkler

system; alternatively activation may be by an automatic fire detection system which conforms to BS 5839-1:2002 (at least L3 standard). For further information on equipment for removing hot smoke refer to BS EN 12101-3:2002.

8.5.3   Construction of outlet ducts or shafts

Outlet ducts or shafts, including any bulkheads over them (see figure 47), should be enclosed in non-combustible construction having not less fire resistance than the element through which they pass.

Where there are natural smoke outlet shafts from different compartments of the same basement storey, or from different basement storeys, they should be separated from each other by non-combustible construction having not less fire resistance than the storey(s) they serve.

Appendices

Appendix A
Performance of materials, products and structures

Introduction

Much of the guidance in this document is given in terms of performance in relation to British or European Standards for products or methods of test or design or in terms of European Technical Approvals. In such cases the material, product or structure should:

a. be in accordance with a specification or design which has been shown by test to be capable of meeting that performance; or

   Note: For this purpose, laboratories accredited by the United Kingdom Accreditation Service (UKAS) for conducting the relevant tests would be expected to have the necessary expertise.

b. have been assessed from test evidence against appropriate standards, or by using relevant design guides, as meeting that performance; or

   Note: For this purpose, laboratories accredited by UKAS for conducting the relevant tests and suitably qualified fire safety engineers might be expected to have the necessary expertise.

   For materials/products where European standards or approvals are not yet available and for a transition period after they become available, British standards may continue to be used. Any body notified to the UK Government by the Government of another member state of the European Union as capable of assessing such materials/products against the relevant British Standards, may also be expected to have the necessary expertise. Where European materials/products standards or approvals are available, any body notified to the European Commission as competent to assess such materials or products against the relevant European standards or technical approval can be considered to have the appropriate expertise.

c. where tables of notional performance are included in this document, conform with an appropriate specification given in these tables; or

d. in the case of fire-resisting elements:

   i. conform with an appropriate specification given in Part II of the Building Research Establishment's report *Guidelines for the construction of fire-resisting structural elements* (BR 128, BRE 1988); or

   ii. be designed in accordance with a relevant British Standard or Eurocode.

   Note 1: Different forms of construction can present different problems and opportunities for the provision of structural fire protection. Further information on some specific forms of construction can be found in;

   Timber – BRE 454 *Multi-storey timber frame buildings – a design guide 2003* (ISBN: 1 86081 605 3).

   Steel – SCI P 97 *Designing for structural fire safety: A handbook for architects and engineers 1999* (ISBN: 1 85942 074 5).

   Note 2: Any test evidence used to substantiate the fire resistance rating of a construction should be carefully checked to ensure that it demonstrates compliance that is adequate and applicable to the intended use. Small differences in detail (such as fixing method, joints, dimensions and the introduction of insulation materials, etc) may significantly affect the rating.

Building Regulations deal with fire safety in buildings as a whole. Thus they are aimed at limiting fire hazard. The aim of standard fire tests is to measure or assess the response of a material, product, structure or system to one or more aspects of fire behaviour. Standard fire tests cannot normally measure fire hazard. They form only one of a number of factors that need to be taken into account. Other factors are set out in this publication.

Fire resistance

Factors having a bearing on fire resistance, that are considered in this document, are:

a. fire severity;

b. building height; and

c. building occupancy.

The standards of fire resistance given are based on assumptions about the severity of fires and the consequences should an element fail. Fire severity is estimated in very broad terms from the use of the building (its purpose group), on the assumption that the building contents (which constitute the fire load) are similar for buildings in the same use.

A number of factors affect the standard of fire resistance specified. These are:

a. the amount of combustible material per unit of floor area in various types of building (the fire load density);

b. the height of the top floor above ground, which affects the ease of escape and of fire-fighting operations and the consequences should large scale collapse occur;

c. occupancy type, which reflects the ease with which the building can be evacuated quickly;

d. whether there are basements, because the lack of an external wall through which to vent heat and smoke may increase heat build-up and thus affect the duration of a fire, as well as complicating fire-fighting; and

e. whether the building is of single storey construction (where escape is direct and structural failure is unlikely to precede evacuation).

Because the use of buildings may change, a precise estimate of fire severity based on the fire load due to a particular use may be misleading. Therefore if a fire engineering approach of this kind is adopted the likelihood that the fire load may change in the future needs to be considered.

Performance in terms of the fire resistance to be met by elements of structure, doors and other forms of construction is determined by reference to either:

a. (National tests) BS 476 *Fire tests on building materials and structures*, Parts 20-24: 1987, ie, Part 20 *Method for determination of the fire resistance of elements of construction (general principles)*, Part 21 *Methods for determination of the fire resistance of loadbearing elements of construction*, Part 22 *Methods for determination of the fire resistance of non-loadbearing elements of construction*, Part 23 *Methods for determination of the contribution of components to the fire resistance of a structure* and Part 24 *Method for determination of the fire resistance of ventilation ducts* (or to BS 476-8:1972 in respect of items tested or assessed prior to January 1988); or

b. (European tests) Commission Decision 2000/367/EC of 3rd May 2000 implementing Council Directive 89/106/EEC as regards the classification of the resistance to fire performance of construction products, construction works and parts thereof.

Note: The designation of xxxx is used for the year reference for standards that are not yet published. The latest version of any standard may be used provided that it continues to address the relevant requirements of the Regulations.

All products are classified in accordance with BS EN 13501-2:2003, *Fire classification of construction products and building elements – Classification using data from fire resistance tests (excluding products for use in ventilation systems).*

BS EN 13501-3:2005, *Fire classification of construction products and building elements – Classification using data from fire resistance tests on components of normal building service installations (other than smoke control systems).*

BS EN 13501-4:xxxx, *Fire classification of construction products and building elements – Classification using data from fire resistance tests on smoke control systems.*

The relevant European test methods under BS EN 1364, 1365, 1366 and 1634 are listed in appendix H of Approved Document B.

Table A1 gives the specific requirements for each element in terms of one or more of the following performance criteria:

a. resistance to collapse (loadbearing capacity), which applies to loadbearing elements only, denoted R in the European classification of the resistance to fire performance;

b. resistance to fire penetration (integrity), denoted E in the European classification of the resistance to fire performance; and

c. resistance to the transfer of excessive heat (insulation), denoted I in the European classification of the resistance to fire performance.

Table A2 sets out the minimum periods of fire resistance for elements of structure.

Table A3 sets out criteria appropriate to the suspended ceilings that can be accepted as contributing to the fire resistance of a floor.

Table A4 sets out limitations on the use of uninsulated fire-resisting glazed elements.

These limitations do not apply to the use of insulated fire-resisting glazed elements.

Information on tested elements is frequently given in literature available from manufacturers and trade associations.

Information on tests on fire-resisting elements is also given in such publications as:

Association for Specialist Fire Protection, *Fire protection for structural steel in buildings 4th Edition* (ISBN: 1 87040 925 6).

Roofs

Performance in terms of the resistance of roofs to external fire exposure is determined by reference to either:

a. (National tests) BS 476-3:2004 *External fire exposure roof tests*; or

b. (European tests) Commission Decision XXXX/YYYY/EC amending Decision 2001/671/EC establishing a classification system for the external fire performance of roofs and roof coverings.

Constructions are classified within the National system by 2 letters in the range A to D, with an AA designation being the best. The first letter indicates the time to penetration; the second letter a measure of the spread of flame.

Constructions are classified within the European system as $B_{ROOF}(t4)$, $C_{ROOF}(t4)$, $D_{ROOF}(t4)$, $E_{ROOF}(t4)$ or $F_{ROOF}(t4)$ (with $B_{ROOF}(t4)$ being the highest performance and $F_{ROOF}(t4)$ being the lowest) in accordance with BS EN 13501-5:2005, *Fire classification of construction products and building elements – Classification using test data from external fire exposure to roof tests.*

BS EN 13501-1 refers to four separate tests. The suffix (t4) used above indicates that Test 4 is to be used for the purposes of this document.

Some roof covering products (and/or materials) can be considered to fulfil all of the requirements for the performance characteristic 'external fire performance' without the need for testing, subject to any national provisions on the design and execution of works being fulfilled. These roof covering products are listed in Commission Decision 2000/553/EC of 6th September 2000 implementing Council Directive 89/106/EEC as regards the external fire performance of roof coverings.

In some circumstances roofs, or parts of roofs, may need to be fire-resisting, for example if used as an escape route or if the roof performs the function of a floor. Such circumstances are covered in sections 4.4 and 5.2.

Table A5 gives notional designations of some generic roof coverings.

Reaction to fire

Performance in terms of reaction to fire to be met by construction products is determined by Commission Decision 200/147/EC of 8th February 2000 implementing Council Directive 89/106/EEC as regards the classification of the reaction to fire performance of construction products.

Note: The designation of xxxx is used for the year reference for standards that are not yet published. The latest version of any standard may be used provided that it continues to address the relevant requirements of the Regulations.

All products, excluding floorings, are classified as A1, A2, B, C, D, E or F (with class A1 being the highest performance and F being the lowest) in accordance with BS EN 13501-1:2002, *Fire classification of construction products and building elements, Part 1 – Classification using data from reaction to fire tests.*

The classes of reaction to fire performance of A2, B, C, D and E are accompanied by additional classifications related to the production of smoke (s1, s2, s3) and/or flaming droplets/particles (d0, d1, d2).

The relevant European test methods are specified as follows,

BS EN ISO 1182:2002, *Reaction to fire tests for building products – Non-combustibility test.*

BS EN ISO 1716:2002, *Reaction to fire tests for building products – Determination of the gross calorific value.*

BS EN 13823:2002, *Reaction to fire tests for building products – Building products excluding floorings exposed to the thermal attack by a single burning item.*

BS EN ISO 925-2:2002, *Reaction to fire tests for building Products, Part 2 – Ignitability when subjected to direct impingement of a flame.*

BS EN 13238:2001, *Reaction to fire tests for building products – conditioning procedures and general rules for selection of substrates.*

Non-combustible materials

Non-combustible materials are defined in table A6 either as listed products, or in terms of performance:

a. (National classes) when tested to BS 476-4:1970 *Non-combustibility test for materials* or BS 476-11:1982 *Method for assessing the heat emission from building products*; or

b. (European classes) when classified as class A1 in accordance with BS EN 13501-1:2002, *Fire classification of construction products and building elements, Part 1 – Classification using data from reaction to fire tests when tested* to BS EN ISO 1182:2002, *Reaction to fire tests for building products – Non-combustibility test* and BS EN ISO 1716:2002 *Reaction to fire tests for building products – Determination of the gross calorific value.*

Table A6 identifies non-combustible products and materials and lists circumstances where their use is necessary.

Materials of limited combustibility

Materials of limited combustibility are defined in table A7:

a. (National classes) by reference to the method specified in BS 476: Part 11:1982; or

b. (European classes) in terms of performance when classified as class A2-s3, d2 in accordance with BS EN 13501-1:2002, *Fire classification of construction products and building elements, Part 1 – Classification using data from reaction to fire tests when tested* to BS EN ISO 1182:2002, *Reaction to fire tests for building products – Non-combustibility test* or BS EN ISO 1716:2002 *Reaction to fire tests for building products – Determination of the gross calorific value* and BS EN 13823:2002, *Reaction to fire tests for building products – Building products excluding floorings exposed to the thermal attack by a single burning item.*

Table A7 also includes composite products (such as plasterboard) which are considered acceptable and where these are exposed as linings they should also meet any appropriate flame spread rating.

Internal linings

Flame spread over wall or ceiling surfaces is controlled by providing for the lining materials or products to meet given performance levels in tests appropriate to the materials or products involved.

Under the National classifications, lining systems which can be effectively tested for 'surface spread of flame' are rated for performance by reference to the method specified in BS 476-7:1971 *Surface spread of flame tests for materials*, or 1987 *Method for classification of the surface spread of flame of products*, or 1997 *Method of test to determine the classification of the surface spread of flame of products* under which materials or products are classified 1, 2, 3 or 4 with Class 1 being the highest.

Under the European classifications, lining systems are classified in accordance with BS EN

13501-1:2002, *Fire classification of construction products and building elements, Part 1– Classification using data from reaction to fire tests.*

Materials or products are classified as A1, A2, B, C, D, E or F, with A1 being the highest. When a classification includes 's3, d2', it means that there is no limit set for smoke production and/or flaming droplets/particles.

To restrict the use of materials which ignite easily, which have a high rate of heat release and/or which reduce the time to flashover, maximum acceptable 'fire propagation' indices are specified, where the National test methods are being followed. These are determined by reference to the method specified in BS 476-6:1981 or 1989 *Method of test for fire propagation of products.* Index of performance (I) relates to the overall test performance, whereas sub-index (i1) is derived from the first three minutes of test.

The highest National product performance classification for lining materials is Class 0. This is achieved if a material or the surface of a composite product is either:

a. composed throughout of materials of limited combustibility; or

b. a Class 1 material which has a fire propagation index (I) of not more than 12 and sub-index (i1) of not more than 6.

Note: Class 0 is not a classification identified in any British Standard test.

Composite products defined as materials of limited combustibility (see 'materials of limited combustibility' above and table A7) should in addition comply with the test requirement appropriate to any surface rating specified in the guidance on requirements B2, B3 and B4.

The notional performance ratings of certain widely used generic materials or products are listed in table A8 in terms of their performance in the traditional lining tests BS 476 Parts 6 and 7 or in accordance with BS EN 13501-1:2002, *Fire classification of construction products and building elements, Part 1– Classification using data from reaction to fire tests.*

Results of tests on proprietary materials are frequently given in literature available from manufacturers and trade associations.

Any reference used to substantiate the surface spread of flame rating of a material or product should be carefully checked to ensure that it is suitable, adequate and applicable to the construction to be used. Small differences in detail, such as thickness, substrate, colour, form, fixings, adhesive, etc, may significantly affect the rating.

Thermoplastic materials

A thermoplastic material means any synthetic polymeric material which has a softening point below 200°C if tested to BS EN ISO 306:2004 method A120 *Plastics – Thermoplastic materials – Determination of Vicat softening temperature.* Specimens for this test may be fabricated from the original polymer where the thickness of material of the end product is less than 2.5mm.

A thermoplastic material in isolation can not be assumed to protect a substrate, when used as a lining to a wall or ceiling. The surface rating of both products must therefore meet the required classification. If however, the thermoplastic material is fully bonded to a non-thermoplastic substrate, then only the surface rating of the composite will need to comply. Concessions are made for thermoplastic materials used for window glazing, rooflights and lighting diffusers within suspended ceilings, which may not comply with the criteria specified in 'internal linings' onwards. They are described in the guidance on requirements B2 and B4.

For the purposes of the requirements B2 and B4 thermoplastic materials should either be used according to their classification 0-3, under the BS 476: Parts 6 and 7 tests as described in 'internal linings' onwards, (if they have such a rating), or they may be classified TP(a) rigid, TP(a) flexible, or TP(b) according to the following methods:

TP(a) rigid:

i.   rigid solid PVC sheet;

ii. solid (as distinct from double- or multiple-skin) polycarbonate sheet at least 3mm thick;

iii. multi-skinned rigid sheet made from unplasticised PVC or polycarbonate which has a Class 1 rating when tested to BS 476-7:1971, 1987 or 1997; or

iv. any other rigid thermoplastic product, a specimen of which (at the thickness of the product as put on the market), when tested to BS 2782-0:2004 Method 508A *Rate of burning, Laboratory method*, performs so that the test flame extinguishes before the first mark and the duration of flaming or afterglow does not exceed 5 seconds following removal of the burner.

TP(a) flexible:

Flexible products not more than 1mm thick which comply with the Type C requirements of *BS 5867-2:1980 Specification for fabrics for curtains and drapes – Flammability requirements when tested* to BS 5438:1989.

Methods of test for flammability of textile fabrics when subjected to a small igniting flame applied to the face or bottom edge of vertically oriented specimens.

Test 2, with the flame applied to the surface of the specimens for 5, 15, 20 and 30 seconds respectively, but excluding the cleansing procedure; and

TP(b):

i.   rigid solid polycarbonate sheet products less than 3mm thick, or multiple-skin polycarbonate sheet products which do not qualify as TP(a) by test; or

ii. other products which, when a specimen of the material between 1.5 and 3mm thick is tested in accordance with BS2782-0:2004 Method 508A, has a rate of burning which does not exceed 50mm/minute.

Note: If it is not possible to cut or machine a 3mm thick specimen from the product then a 3mm test specimen can be moulded from the same material as that used for the manufacture of the product.

Note: Currently, no new guidance is possible on the assessment or classification of thermoplastic materials under the European system since there is no generally accepted European test procedure and supporting comparative data.

Fire test methods

A guide to the various test methods in BS 476 and BS 2782 is given in PD 6520: *Guide to fire test methods for building materials and elements of construction* (available from the British Standards Institution).

A guide to the development and presentation of fire tests and their use in hazard assessment is given in BS 6336:1998 *Guide to development and presentation of fire tests and their use in hazard assessment.*

Table A1 Specific provisions of test for fire resistance of elements of structure etc					
Part of building	Minimum provisions when tested to the relevant part of BS 476[1] (minutes)			Minimum provisions when tested to the relevant European standard (minutes)[9]	Method of exposure
	Loadbearing capacity[2]	Integrity	Insulation		
1. Structural frame, beam or column.	See table A2	Not applicable	Not applicable	R see table A2	Exposed faces
2. Loadbearing wall (which is not also a wall described in any of the following items).	See table A2	Not applicable	Not applicable	R see table A2	Each side separately

Table A1 continued on next page

Table A1 Specific provisions of test for fire resistance of elements of structure etc					
3. Floors[3] Any floor – including compartment floors	See table A2	See table A2	See table A2	REI see table A2	From underside[4]
4. Roofs					
a. Any part forming an escape route	30	30	30	REI 30	From underside[4]
b. Any roof that performs the function of a floor	See table A2	See table A2	See table A2	REI see table A2	From underside[4]
5. External walls					
a. Any part less than 1000mm from any point on the relevant boundary[5]	See table A2	See table A2	See table A2	REI see table A2	Each side separately
b. Any part 1000mm or more from the relevant boundary[5]	See table A2	See table A2	15	RE see table A2 and REI 15	From inside the building
c. Any part adjacent to an external escape route (see section 4.4.8, figure 23)	30	30	No provision[6][7]	RE 30	From inside the building
6. Compartment walls	See table A2	See table A2	See table A2	REI see table A2	Each side separately
7. Protected shafts excluding any fire-fighting shaft					
a. Any glazing described in section 6.3.5.3	Not applicable	30	No provision[7]	E 30	Each side separately
b. Any other part between the shaft and a protected lobby/corridor described in section 6.3.5.3	30	30	30	REI 30	Each side separately
c. Any part not described in (a) or (b) above.	See table A2	See table A2	See table A2	REI see table A2	Each side separately

Table A1 Specific provisions of test for fire resistance of elements of structure etc

8. Enclosure (which does not form part of a compartment wall or a protected shaft) to a:					
a. protected stairway	30	30	30[8]	REI 30[8]	Each side separately
b. lift shaft	30	30	30	REI 30	Each side separately
9. Fire-fighting shafts	120	120	120	REI 120	From side remote from shaft
a. Construction separating fire-fighting shaft from rest of building	60	60	60	REI 60	From shaft side
b. Construction separating fire-fighting stair, fire-fighting lift shaft and fire-fighting lobby	60	60	60	REI 60	Each side separately
10. Enclosure (which is not a compartment wall or described in item 7) to a:					
a. protected lobby	30	30	30[8]	REI 30[8]	Each side separately
b. protected corridor	30	30	30[8]	REI 30[8]	Each side separately
11. Sub-division of a corridor	30	30	30[8]	REI 30[8]	Each side separately
12. Fire-resisting construction:					
a. enclosing places of special fire risk (see section 6.3.2.2)[10]	30	30	30	REI 30	Each side separately
b. fire-resisting subdivision described in section 4.3.2.16, figure 18(b)	30	30	30	REI 30	Each side separately
13. Cavity barrier [11]	Not applicable	30	15	E 30 and EI 15	Each side separately
14. Ceiling (figure 32)	Not applicable	30	30	EI 30	From underside

Table A1 continued on next page

Table A1 Specific provisions of test for fire resistance of elements of structure etc

15. Duct described in section 6.4.5	Not applicable	30	No provision	E 30	From outside
16. Casing around a drainage system described in section 6.5, figure 35	Not applicable	30	No provision	E 30	From outside
17. Flue walls described in section 6.5, figure 36	Not applicable	Half the period specified in table A2 for the compartment wall/floor	Half the period specified in table A2 for the compartment wall/floor	EI half the period specified in table A2 for the compartment wall/floor	From outside
18. Fire doors	See table C1	See table C1	See table C1	See table C1	

Notes:

1. Part 21 for loadbearing elements, Part 22 for non-loadbearing elements, Part 23 for fire-protecting suspended ceilings, and Part 24 for ventilation ducts. BS 476-8 results are acceptable for items tested or assessed before 1 January 1988.

2. Applies to loadbearing elements only (section 6.1.3 and appendix F).

3. Guidance on increasing the fire resistance of existing timber floors is given in BRE Digest 208 *Increasing the fire resistance of existing timber floors* (BRE 1988).

4. A suspended ceiling should only be relied on to contribute to the fire resistance of the floor if the ceiling meets the appropriate provisions given in table A3.

5. The guidance in section 7.2 allows such walls to contain areas which need not be fire-resisting (unprotected areas).

6. Unless needed as part of a wall in item 5a or 5b.

7. Except for any limitations on glazed elements given in table A4.

8. See table A4 for permitted extent of uninsulated glazed elements.

9. The National classifications do not automatically equate with the equivalent classifications in the European column, therefore products cannot typically assume a European class unless they have been tested accordingly.

10. It is recommended that all rooms of special fire hazard should have 60 minutes fire resistance, for property protection purposes.

11. For property protection, cavity barriers should also have 30 minutes insulation fire resistance (rather than 15min) unless:
    - there are no combustible materials used in the construction of the building within 3m of the barrier; and
    - it is impossible or highly unlikely that combustible materials will at any time be stored within 3m of either side of the barrier. If flammable materials may be present, fully-insulating (30min) barriers shall always be used.

12. The guidance provided in Tables A1 and A4 is for the prime purpose of life safety. Levels of fire resistance for protection of the building fabric and its contents may need to be modified accordingly, either for longer resistance times or higher levels of performance (eg, insulation as well as integrity). This may be appropriate to provide enhanced resilience of the structure against the effects of fire over longer exposure times in post flashover conditions than are routinely considered normal for the provision of safe escape.

'R' is the European classification of the resistance to fire performance in respect of loadbearing capacity; 'E' is the European classification of the resistance to fire performance in respect of integrity; and 'I' is the European classification of the resistance to fire performance in respect of insulation.

Table A2 Minimum periods of fire resistance for school buildings[1][5]

	Minimum periods of fire resistance (minutes) in a:		
	Basement storey (including floor over), not more than 10m deep[2]	Ground or upper storey; Height (m) of top floor above ground, in a building or separated part of a building	
		Not more than 5	Not more than 18[3]
Not sprinklered	60	60	60
Sprinklered	60	30[4]	60

Notes

1. For buildings with other purpose groups, refer to appendix A of Approved Document B.

2. For buildings with basements more than 10m deep, refer to appendix A of Approved Document B

3. For buildings with a top floor more than 18m above ground, refer to appendix A of Approved Document B.

4. Increased to a minimum of 60 minutes for compartment walls separating buildings.

5. For property protection, add 60 minutes to the above periods if the compartment wall or floor separates two different occupancies.

Application of the fire resistance standards in table A2:

a. Where one element of structure supports or carries or gives stability to another, the fire resistance of the supporting element should be no less than the minimum period of fire resistance for the other element (whether that other element is loadbearing or not).

There are circumstances where it may be reasonable to vary this principle, for example:

i. where the supporting structure is in the open air and is not likely to be affected by the fire in the building; or

ii. the supporting structure is in a different compartment, with a fire-separating element (which has the higher standard of fire resistance) between the supporting and the separated structure; or

iii. where a plant room on the roof needs a higher fire resistance than the elements of structure supporting it.

b. Where an element of structure forms part of more than one building or compartment, that element should be constructed to the standard of the greater of the relevant provisions.

c. Where one side of a basement is (due to the slope of the ground) open at ground level, giving an opportunity for smoke venting and access for fire-fighting, it may be appropriate to adopt the standard of fire resistance applicable to above-ground structures for elements of structure in that storey.

d. Although most elements of structure in a single storey building may not need fire resistance (see the guidance on requirement B3, section 6.2.2.2), fire resistance will be needed if the element:

i. is part of (or supports) an external wall and there is provision in the guidance on requirement B4 to limit the extent of openings and other unprotected areas in the wall; or

ii. is part of (or supports) a compartment wall, including a wall common to two or more buildings; or

iii. supports a gallery.

For the purposes of point 'd' above, the ground storey of a building which has one or more basement storeys and no upper storeys, may be considered as a single storey building. The fire resistance of the basement storeys should be that appropriate to basements.

Table A3 Limitations on fire-protecting suspended ceilings (see table A1, note 4)

Height of building or separated part (m)	Type of floor	Provision for fire resistance of floor (minutes)	Description of suspended ceiling
Less than 18	Not compartment	60 or less	Type W, X, Y or Z
	Compartment	less than 60	
		60	Type X, Y or Z
18 or more	any	60 or less	Type Y or Z
No limit	any	More than 60	Type Z

Notes:

1. Ceiling type and description (the change from Types A-D to Types W-Z is to avoid confusion with Classes A-D (European)):

   W. Surface of ceiling exposed to the cavity should be Class 0 or Class 1 (National) or Class C-s3, d2 or better (European).

   X. Surface of ceiling exposed to the cavity should be Class 0 (National) or Class B-s3, d2 or better (European).

   Y. Surface of ceiling exposed to the cavity should be Class 0 (National) or Class B-s3, d2 or better (European). Ceiling should not contain easily openable access panels.

   Z. Ceiling should be of a material of limited combustibility (National) or of Class A2-s3, d2 or better (European) and not contain easily openable access panels. Any insulation above the ceiling should be of a material of limited combustibility (National) or Class A2-s3, d2 or better (European).

2. Any access panels provided in fire protecting suspended ceilings of type Y or Z should be secured in position by releasing devices or screw fixings, and they should be shown to have been tested in the ceiling assembly in which they are incorporated.

3. The National classifications do not automatically equate with the equivalent European classifications, therefore, products cannot typically assume a European class unless they have been tested accordingly.

   When a classification includes 's3, d2', this means that there is no limit set for smoke production and/or flaming droplets/particles.

Table A4 Limitations on the use of uninsulated glazed elements on escape routes
(These limitations do not apply to glazed elements which satisfy the relevant insulation criterion, see table A1)

Position of glazed element	Maximum total glazed area in parts of a building with access to:			
	A single stairway		More than one stairway	
	Walls	Door leaf	Walls	Door leaf
1. Between a protected stairway[1] and: a. the accommodation; or b. a corridor which is not a protected corridor.	Nil	25% of door area	Unlimited above 1100mm[2]	50% of door area
2. Between: a. a protected stairway[1] and a protected lobby or protected corridor; or b. accommodation and a protected lobby.	Unlimited above 1100mm from floor	Unlimited above 100mm from floor	Unlimited above 100mm from floor	Unlimited above 100mm from floor
3. Between the accommodation and a protected corridor forming a dead end	Unlimited above 1100mm from floor	Unlimited above 100mm from floor	Unlimited above 1100mm from floor	Unlimited above 100mm from floor
4. Between accommodation and any other corridor; or subdividing corridors	Not applicable	Not applicable	Unlimited above 100mm from floor	Unlimited above 100mm from floor
5. Adjacent an external escape route described in section 4.3.2.17	Unlimited above 1100mm from floor	Unlimited above 1100mm from floor	Unlimited above 1100mm from floor	Unlimited above 1100mm from floor
6. Adjacent an external escape stair (see section 4.4.8 and figure 23) or roof escape (see section 4.3.2.18)	Unlimited	Unlimited	Unlimited	Unlimited

Notes:

1. If the protected stairway is also a protected shaft (see section 6.3.5) or a fire-fighting stair (see section 8.4) there may be further restrictions on the uses of glazed elements.

2. Measured vertically from the landing floor level or the stair pitch line.

3. The 100mm limit is intended to reduce the risk of fire spread from a floor covering.
   For property protection, it is recommended to increase the limit to 500mm (NB. AD M provides guidance that vision panels in doors should have a lower edge not more than 500mm above the floor).

4. Items 1 and 6 apply also to single storey buildings.

5. Fire-resisting glass should be marked with the manufacturer and product name.

6. Further guidance can be found in *A guide to best practice in the specification and use of fire-resistant glazed systems* published by the Glass and Glazing Federation.

7. The guidance provided in Tables A1 and A4 is for the prime purpose of life safety. Levels of fire resistance for protection of the building fabric and its contents may need to be modified accordingly, either for longer resistance times or higher levels of performance (eg, insulation as well as integrity). This may be appropriate to provide enhanced resilience of the structure against the effects of fire over longer exposure times in post flashover conditions than are routinely considered normal for the provision of safe escape.

Table A5 Notional designations of roof coverings
Part 1. Pitched roofs covered with slates or tiles

Covering material	Supporting structure	Designation
1. Natural slates	Timber rafters with or without underfelt, sarking, boarding, woodwool slabs, compressed straw slabs, plywood, wood chipboard, or fibre insulating board	AA (National class) or B_{ROOF}(t4) (European class)
2. Fibre reinforced cement slates		
3. Clay tiles		
4. Concrete tiles		

Note: Although the table does not include guidance for roofs covered with bitumen felt, it should be noted that there is a wide range of materials on the market and information on specific products is readily available from manufacturers.

Table A5 Notional designations of roof coverings
Part 2. Pitched roofs covered with self-supporting sheet

Roof covering material	Construction	Supporting structure	Designation
1. Profiled sheet of galvanised steel, aluminium, fibre reinforced cement, or pre-painted (coil coated) steel or aluminium with PVC or PVF2 coating	Single skin without underlay, or with underlay or plasterboard, fibre insulating board or woodwool slab	Structure of timber, steel or concrete	AA (National class) or B_{ROOF}(t4) (European class)
2. Profiled sheet of galvanised steel, aluminium, fibre reinforced cement, or pre-painted (coil coated) steel or aluminium with PVC or PVF2 coating	Double skin without interlayer or with interlayer of resin bonded glass fibre, mineral wool slab, polystyrene or polyurethane	Structure of timber, steel or concrete	AA (National class) or B_{ROOF}(t4) (European class)

Table A5 Notional designations of roof coverings
Part 3. Flat roofs covered with bitumen felt

A flat roof comprising of bitumen felt should (irrespective of the felt specification) be deemed designation AA (National class) or B_{ROOF}(t4) (European class) if the felt is laid on a deck constructed of 6mm plywood, 12.5mm wood chipboard, 16mm (finished) plain edged timber boarding, compressed straw slab, screeded woodwool slab, profiled fibre reinforced cement or steel deck (single or double skin) with or without fibre insulating board overlay, profiled aluminium deck (single or double skin) with or without fibre insulating board overlay, or concrete or clay pot slab (in situ or precast) and has a finish of:
a. bitumen-bedded stone chippings covering the whole surface to a depth of at least 12.5mm;
b. bitumen-bedded tiles of a non-combustible material;
c. sand and cement screed; or
d. macadam.

Table A5 Notional designations of roof coverings
Part 4. Pitched or flat roofs covered by fully supported material

Covering material	Supporting structure	Designation
1. Aluminium sheet	Timber joists and tongued and grooved boarding or plain edged boarding	AA* (National class) or $B_{ROOF}(t4)$ (European class)
2. Copper sheet		
3. Zinc sheet		
4. Lead sheet		
5. Mastic asphalt		
6. Vitreous enamelled steel	Steel or timber joists with deck of: woodwool slabs, compressed straw slab, wood chipboard, fibre insulating board, or 9.5mm plywood.	AA (National class) or $B_{ROOF}(t4)$ (European class)
7. Lead/tin alloy coated steel sheet		
8. Zinc/aluminium alloy coated steel sheet		
9. Pre-painted (coil coated) steel sheet including liquid-applied PVC coatings	Concrete or clay pot slab (in situ or precast) or non-combustible deck of steel, aluminium or fibre cement (with or without insulation)	AA (National class) or $B_{ROOF}(t4)$ (European class)

Notes: (*) Lead sheet supported by timber joists and plain edged boarding should be regarded as having a BA designation and is deemed to be designated $C_{ROOF}(t4)$ (European class).

The National classifications do not automatically equate with the equivalent classifications in the European column; therefore, products cannot typically assume a European class unless they have been tested accordingly.

Table A6 Use and definitions of non-combustible materials

References in BB 100 guidance to situations where such materials should be used	Definitions of non-combustible materials	
	National class	European class
1. refuse chutes meeting the provisions in the guidance to B3, section 6.3.4.3

2. suspended ceilings and their supports where there is provision in the guidance to B3, section 6.4.4.1, for them to be constructed of non-combustible materials

3. pipes meeting the provisions in the guidance to B3, table 11

4. flue walls meeting the provisions in the guidance to B3, figure 36 | a. Any material which when tested to BS 476- 11: 1982 does not flame nor cause any rise in temperature on either the centre (specimen) or furnace thermocouples.
b. Totally inorganic materials such as concrete, fired clay, ceramics, metals, plaster and masonry containing not more than 1% by weight or volume of organic material. (Use in buildings of combustible metals such as magnesium/aluminium alloys should be assessed in each individual case).
c. Concrete bricks or blocks meeting BS EN 771-1 :2003
d. Products classified as non-combustible under BS 476-4:1970. | a. Any material classified as class A1 in accordance with BS EN 13501-1:2002. *Fire classification of construction products and building elements, Part 1 – Classification using data from reaction to fire tests.*
b. Products made from one or more of the materials considered as Class A1 without the need for testing as defined in Commission Decision 2003/424/EC of 6th June 2003 amending Decision 96/603/EC establishing the list of products belonging to Class A1 'No contribution to fire' provided for in the Decision 94/611/EC implementing Article 20 of the Council Directive 89/106/EEC on construction products. None of the materials shall contain more than 1% by weight or volume (whichever is the more onerous) of homogeneously distributed organic material. |

Note: The National classifications do not automatically equate with the equivalent classifications in the European column, therefore products cannot typically assume a European class unless they have been tested accordingly.

Table A7 Use and definitions of materials of limited combustibility

References in BB 100 guidance to situations where such materials should be used	Definitions of non-combustible materials	
	National class	European class

1. Stairs where there is provision in the guidance to B1 for them to be constructed of materials of limited combustibility (see section 4.5.4.1)
 	
2. Materials above a suspended ceiling meeting the provisions in the guidance to B3, section 6.4.4.1
 	
3. Reinforcement/support for fire-stopping referred to in the guidance to B3, see section 6.5.4
 	
4. Roof coverings meeting provisions:
  a. in the guidance to B3, section 6.3.3.6; or
  b. in the guidance to B4, table 15; or
  c. in the guidance to B4, figure 42
 	
5. Roof deck meeting the provisions of the guidance to B3, figure 28a
 	
6. Class 0 materials meeting the provisions in appendix A
 	
7. Ceiling tiles or panels of any fire protecting suspended ceiling (Type Z) in table A3
	a. Any non-combustible material listed in table A6.
	
	b. Any material of density 300 kg/m^3 or more, which when tested to BS 476-11:1982, does not flame and the rise in temperature on the furnace thermocouple is not more than 20°C.
	
	c. Any material with a non-combustible core at least 8mm thick having combustible facings (on one or both sides) not more than 0.5mm thick. (Where a flame spread rating is specified, these materials must also meet the appropriate test requirements).
		a. Any material listed in table A6
		b. Any material/product classified as Class A2-s3, d2 or better in accordance with BS EN 13501-1:2002 Fire classification of construction products and building elements, Part 1 – Classification using data from reaction to fire tests

8. Insulation above any fire-protecting suspended ceiling (Type Z) in table A3 9. Storage lockers	Any of the materials (a), (b) or (c) above, or: d. Any material of density less than 300kg/m^3, which when tested to BS 476-11:1982, does not flame for more than 10 seconds and the rise in temperature on the centre (specimen) thermocouple is not more than 35°C and on the furnace thermocouple is not more than 25°C.	Any of the materials/products (a) or (b) above.

Notes:

1. The National classifications do not automatically equate with the equivalent classifications in the European column; therefore, products cannot typically assume a European class unless they have been tested accordingly.

2. When a classification includes 's3, d2', this means that there is no limit set for smoke production and/or flaming droplets/particles.

Table A8 Typical performance ratings of some generic materials and products

Rating	Material or product
Class 0 (National)	1. Any non-combustible material or material of limited combustibility. (Composite products listed in table A7 must meet test requirements given in appendix A) 2. Brickwork, blockwork, concrete and ceramic tiles. 3. Plasterboard (painted or not with a PVC facing not more than 0.5mm thick) with or without an air gap or fibrous or cellular insulating material behind. 4. Woodwool cement slabs. 5. Mineral fibre tiles or sheets with cement or resin binding.
Class 3 (National)	6. Timber or plywood with a density greater than 400kg/m³, painted or unpainted. 7. Wood particle board or hardboard, either untreated or painted. 8. Standard glass reinforced polyesters.
Class A1 (European)	9. Any material that achieves this class or is defined as 'classified without further test' in a published Commission Decision.
Class A2-s3, d2 (European)	10. Any material that achieves this class or is defined as 'classified without further test' in a published Commission Decision.
Class B-s3, d2 (European)	11. Any material that achieves this class or is defined as 'classified without further test' in a published Commission Decision
Class C-s3, d2 (European)	12. Any material that achieves this class or is defined as 'classified without further test' in a published Commission Decision.
Class D-s3, d2 (European)	13. Any material that achieves this class or is defined as 'classified without further test' in a published Commission Decision.

Notes (National):

1. Materials and products listed under Class 0 also meet Class 1.

2. Timber products listed under Class 3 can be brought up to Class 1 with appropriate proprietary treatments.

3. The following materials and products may achieve the ratings listed below. However, as the properties of different products with the same generic description vary, the ratings of these materials/products should be substantiated by test evidence.

   Class 0: Aluminium faced fibre insulating board, flame retardant decorative laminates on a calcium silicate board, thick polycarbonate sheet, phenolic sheet and UPVC.

   Class 1: Phenolic or melamine laminates on a calcium silicate substrate and flame-retardant decorative laminates on a combustible substrate.

Notes (European):

For the purposes of the Building Regulations:

1. Materials and products listed under Class A1 also meet Classes A2-s3, d2, B-s3, d2, C-s3, d2 and D-s3, d2.

2. Materials and products listed under Class A2-s3, d2 also meet Classes B-s3, d2, C-s3, d2 and D-s3, d2.

3. Materials and products listed under Class B-s3, d2 also meet Classes C-s3, d2 and D-s3, d2.

4. Materials and products listed under Class C-s3, d2 also meet Class D-s3, d2.

5. The performance of timber products listed under Class D-s3, d2 can be improved with appropriate proprietary treatments.

6. Materials covered by the CWFT process (classification without further testing) can be found by accessing the European Commission's website via the link on the CLG website www.communities.gov.uk

7. The national classifications do not automatically equate with the equivalent classifications in the European column, therefore products cannot typically assume a European class unless they have been tested accordingly.

8. When a classification includes 's3, d2', this means that there is no limit set for smoke production and/or flaming droplets/particles.

Insulating core panel systems are used for external cladding as well as for internal structures. However, whilst both types of panel system have unique fire behaviour characteristics, it is those used for internal structures that can present particular problems with regard to fire spread.

These panels typically consist of an inner core sandwiched between and bonded to facings of galvanised steel, often with a PVC facing for hygiene purposes. The panels are then formed into a structure by jointing systems, usually designed to provide an insulating and hygienic performance. The panel structure can be free standing, but is usually attached to the building structure by lightweight fixings or hangers in the case of ceilings.

The most common forms of insulation in present use are:

- polyisocyanurate;
- polyurethane;
- mineral fibre;
- phenolic;
- polystyrene;
- extruded polystyrene; and
- composite polymers such as syntactic phenolic.

Fire behaviour of the core materials and fixing systems

The degradation of polymeric materials can be expected when exposed to radiated/conducted heat from a fire, with the resulting production of large quantities of smoke.

It is recognised that the potential for problems in fires involving mineral fibre cores is generally less than those for polymeric core materials.

In addition, irrespective of the type of core material, the panel, when exposed to the high temperatures of a developed fire, will tend to delaminate between the facing and core material, due to a combination of expansion of the metal facing and softening of the bond line.

Therefore once it is involved, either directly or indirectly in a fire, the panel will have lost most of its structural integrity. Stability will then be dependant on the method of fixing to the structure. For systems that are not fixed through both facings the stability of the system will then depend on the residual structural strength of the non-exposed facing, the interlocking joint between panels and the fixing system.

Most jointing or fixing systems for these systems have an extremely limited structural integrity performance in developed fire conditions. If the fire starts to heat up the support fixings or structure to which they are attached, then there is a real chance of total collapse of the panel system.

Where panels are used as the lining to a building the insulating nature of these panels, together with their sealed joints, means that fire can spread behind the panels, hidden from the occupants of occupied rooms/spaces. With some thermoplastic cores fire can also spread between the panel facings.

This can prove to be as particular problem to fire-fighters as, due to the insulating properties of the cores, it may not be possible to track the spread of fire, even using infra red detection equipment. This difficulty, together with that of controlling the fire spread within and behind the panels, is likely to have a detrimental effect on the performance of the fixing systems, potentially leading to their complete and unexpected collapse, together with any associated equipment.

Fire-fighting

When compared with other types of construction techniques, these panel systems therefore provide a unique combination of problems for fire-fighters, including:

- hidden fire spread within panels with thermoplastic cores;
- production of large quantities of black toxic smoke;
- rapid fire spread leading to flashover; and
- hidden fire behind lining systems.

These four characteristics are common to both polyurethane and polystyrene cored panels, although the rate of fire spread in polyurethane cores is significantly less than that of polystyrene cores, especially when any external heat source is removed.

In addition, irrespective of the type of panel core, all systems are susceptible to:

- delamination of the steel facing;
- collapse of the system; and
- hidden fire spread behind the system.

Design recommendations

To identify the appropriate solution, a risk assessment approach should be adopted. This would involve identifying the potential fire risk within the enclosures formed by the panel systems and then adopting one or more of the following at the design stage:

- removing the risk;
- separating the risk from the panels by an appropriate distance;
- providing a fire suppression system for the risk;
- providing a fire suppression system for the enclosure;
- providing fire-resisting panels; and
- specifying appropriate materials/fixing and jointing systems.

In summary the performance of the building structure, including the insulating envelope, the superstructure, the substructure etc, must be considered in relation to their performance in the event of a fire.

Specifying panel core materials

Where at all possible the specification of panels with core materials appropriate to the application will help ensure an acceptable level of performance for panel systems, when involved in fire conditions.

The following are examples in the provision of core materials which may be appropriate to the application concerned.

Mineral fibre cores:

- cooking areas;
- hot areas;
- fire breaks in combustible panels;
- fire stop panels; and
- general fire protection.

All cores:

- chill stores;
- cold stores; and
- clean rooms.

Note: Core materials may be used in other circumstances where a risk assessment has been made and other appropriate fire precautions have been put in place.

Specifying materials/fixing and jointing systems

The following are methods by which the stability of panel systems may be improved in the event of a fire, although they may not all be appropriate in every case.

In addition the details of construction of the insulating envelope should, particularly in relation to combustible insulant cores, prevent the core materials from becoming exposed to the fire and contributing to the fire load.

a. Insulating envelopes, support systems and supporting structure should be designed to allow the envelope to remain structurally stable by alternative means such as catenary action following failure of the bond line between insulant core and facing materials.

This particularly relates to ceilings and will typically require positive attachment of the lower faces of the insulant panels to supports.

b. The building superstructure, together with any elements providing support to the insulating envelope, should be protected to prevent early collapse of the structure or the envelope.

Note: Irrespective of the type of panel provided, it will remain necessary to ensure that the supplementary support method supporting the panels remains stable for an appropriate time period under fire conditions. It is not practical to fire protect light gauge steel members such as purlins and sheeting rails which provide stability to building superstructures and these may be compromised at an early stage of a fire. Supplementary fire-protected heavier gauge steelwork members could be provided at wider intervals than purlins to provide restraint in the event of a fire.

c. In designated high risk areas, consideration should be given to incorporating non-combustible insulant cored panels into wall and ceiling construction at intervals, or incorporating strips of non-combustible material into specified wall and ceiling panels, in order to provide a barrier to fire propagation through the insulant.

d. Correct detailing of the insulating envelope should ensure that the combustible insulant is fully encapsulated by non-combustible facing materials which remain in place during a fire.

e. The panels should incorporate pre-finished and sealed areas for penetration of services.

General

Generally, panels or panel systems should not be used to support machinery or other permanent loads.

Any cavity created by the arrangement of panels, their supporting structure or other building elements should be provided with suitable cavity barriers.

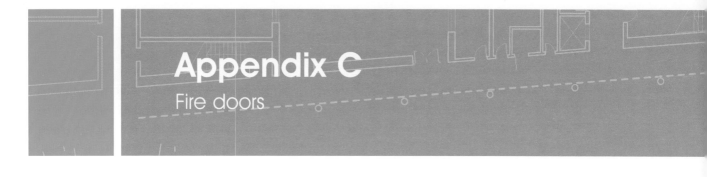

Appendix C
Fire doors

All fire doors should have the appropriate performance given in table C1 either:

a. by their performance under test to BS 476 *Fire tests on building materials and structures*, Part 22 *Methods for determination of the fire resistance of non-loadbearing elements of construction*, in terms of integrity for a period of minutes, eg, FD30. A suffix (S) is added for doors where restricted smoke leakage at ambient temperatures is needed; or

b. as determined with reference to Commission Decision 2000/367/EC of 3rd May 2000 implementing Council Directive 89/106/EEC as regards the classification of the resistance to fire performance of construction products, construction works and parts thereof. All fire doors should be classified in accordance with BS EN 13501-2:xxxx, *Fire classification of construction products and building elements. Classification using data from fire resistance tests (excluding products for use in ventilation systems)*. They are tested to the relevant European method from the following:

- BS EN 1634-1:2000, *Fire resistance tests for door and shutter assemblies. Fire doors and shutters*;

- BS EN 1634-2:xxxx *Fire resistance tests for door and shutter assemblies. Fire door hardware*; and

- BS EN 1634-3:2001 *Fire resistance tests for door and shutter assemblies. Smoke control doors*.

The performance requirement is in terms of integrity (E) for a period of minutes. An additional classification of Sa is used for all doors where restricted smoke leakage at ambient temperatures is needed.

The requirement (in either case) is for test exposure from each side of the door separately, except in the case of lift doors which are tested from the landing side only.

Any test evidence used to substantiate the fire resistance rating of a door or shutter should be carefully checked to ensure that it adequately demonstrates compliance and is applicable to the complete installed assembly. Small differences in detail (such as glazing apertures, intumescent strips, door frames and ironmongery, etc) may significantly affect the rating.

Note 1: The designation of xxxx is used for standards that are not yet published. The latest version of any standard may be used provided that it continues to address the relevant requirements of the Regulations.

Note 2: Until such time that the relevant harmonised product standards are published, for the purposes of meeting the Building Regulations, products tested in accordance with BS EN 1634-1 (with or without pre-fire test mechanical conditioning) will be deemed to have satisfied the provisions provided that they achieve the minimum fire resistance in terms of integrity, as detailed in table C1.

Many insurers recommend that all products (for the protection of openings and services) should be manufactured within the factory production control standards specified in ISO 9002 as a minimum by independently certified firms.

All fire doors should be fitted with a self-closing device except for fire doors to cupboards and to service ducts which are normally kept locked shut.

Note: All rolling shutters should be capable of being opened and closed manually for fire-fighting purposes (see section 8.4.5).

Where a self-closing device would be considered a hindrance to the normal approved use of the building, self-closing fire doors may be held open by:

a. a fusible link (but not if the door is fitted in an opening provided as a means of escape unless it complies with the next paragraph below); or

b. an automatic release mechanism actuated by an automatic fire detection and alarm system; or

c. a door closer delay device.

Two fire doors may be fitted in the same opening so that the total fire resistance is the sum of their individual fire resistances, provided that each door is capable of closing the opening. In such a case, if the opening is provided as a means of escape, both doors should be self-closing, but one of them may be fitted with an automatic self-closing device and be held open by a fusible link if the other door is capable of being easily opened by hand and has at least 30 minutes fire resistance.

Because fire doors often do not provide any significant insulation, there should be some limitation on the proportion of doorway openings in compartment walls. Therefore no more than 25% of the length of a compartment wall should consist of door openings, unless the doors provide both integrity and insulation to the appropriate level (see appendix A, table A2).

Note: Where it is practicable to maintain a clear space on both sides of the doorway, then the above percentage may be greater.

Roller shutters across a means of escape should only be released by a heat sensor, such as a fusible link or electric heat detector, in the immediate vicinity of the door. Closure of shutters in such locations should not be initiated by smoke detectors or a fire alarm system, unless the shutter is also intended to partially descend to form part of a boundary to a smoke reservoir.

Unless shown to be satisfactory when tested as part of a fire door assembly, the essential components of any hinge on which a fire door is hung should be made entirely from materials having a melting point of at least 800°C.

Except for lift entrance/landing doors, all fire doors should be marked with the appropriate fire safety sign complying with BS 5499-5:2002 according to whether the door is:

a. to be kept closed when not in use (Fire door keep shut);

b. to be kept locked when not in use (Fire door keep locked shut); or

c. held open by an automatic release mechanism or free swing device (Automatic fire door keep clear).

Fire doors to cupboards and to service ducts should be marked on the outside; all other fire doors on both sides.

Tables A1 and A2 set out the minimum periods of fire resistance for the elements of structure to which performance of some doors is linked. table A4 sets out limitations on the use of uninsulated glazing in fire doors.

BS 8214:1990 gives recommendations for the specification, design, construction, installation and maintenance of fire doors constructed with non-metallic door leaves.

Guidance on timber fire-resisting doorsets, in relation to the new European test standard, may be found in *Timber Fire-Resisting Doorsets: maintaining performance under the new European test standard,* published by TRADA.

Guidance for metal doors is given in *Code of practice for fire-resisting metal doorsets* published by the DSMA (Door and Shutter Manufacturers' Association) in 1999.

Hardware used on fire doors can significantly affect performance in fire. Notwithstanding the guidance in this document, guidance is available in *Hardware for fire and escape doors* published by the Builders Hardware Industry Federation.

Table C1 Provision for fire doors in school buildings

Position of door	Minimum fire resistance of door in terms of integrity (minutes) when tested to BS 476 part 22[1]	Minimum fire resistance of door in terms of integrity (minutes) when tested to the relevant European Standard BS EN 1634-1
1. In a compartment wall separating buildings.	As for the wall in which the door is fitted, but a minimum of 60	As for the wall in which the door is fitted, but a minimum of 60
2. In a compartment wall:		
a. Enclosing a protected shaft forming a stairway situated wholly or partly above the adjoining ground;	FD 30S[2]	E30 Sa[3]
b. Enclosing a protected shaft forming a stairway not described in (a) above;	Half the period of fire resistance of the wall in which it is fitted, but 30 minimum and with suffix S[2]	Half the period of fire resistance of the wall in which it is fitted, but 30 minimum and with suffix Sa[3]
c. Enclosing a protected shaft forming a lift or service shaft;	Half the period of fire resistance of the wall in which it is fitted, but 30 minimum	Half the period of fire resistance of the wall in which it is fitted, but 30 minimum
d. Not described in (a), (b) or (c) above.	As for the wall it is fitted in, but add S[2] if the door is used for progressive horizontal evacuation under the guidance to B1	As for the wall it is fitted in, but add Sa[3] if the door is used for progressive horizontal evacuation under the guidance to B1
3. In a compartment floor	As for the floor in which it is fitted	As for the floor in which it is fitted
4. Forming part of the enclosures of:		
a. a protected stairway; or	FD 30S[2]	E30 Sa[3]
b. a lift shaft (see section 4.5.6.2); which does not form a protected shaft in 2(a), (b) or (c) above.	FD 30	E30
5. Forming part of the enclosure of:		
a. a protected lobby approach (or protected corridor) to a stairway;	FD 30S[2]	E30 Sa[3]
b. any other protected corridor; or	FD 20 (S)	E20 Sa[3]
c. a protected lobby approach to lift shaft (see section 4.5.6.2)	FD 30S[2]	E30 Sa[3]
6. Affording access to an external escape route	FD 30	E30
7. Sub-dividing:		
a. corridors connecting alternative exits;	FD 20S[2]	E20 Sa[3]
b. dead-end portions of corridors from the remainder of the corridor	FD 20S[2]	E20 Sa[3]
8. Any door within a cavity barrier	FD 30	E30
9. Any door forming part of the enclosure to a place of special fire risk	FD 20	E20

Notes:

1. To BS 476: Part 22 (or BS 476: Part 8 subject to regulatory guidance)

2. Unless pressurization techniques complying with BS 5588: Part 4 *Fire Precautions in the design, construction and use of buildings Code of Practice for smoke control using pressure differentials* are used, these doors should also either:

    (a) have a leakage rate not exceeding 3m³/m/hour (head and jambs only) when tested at 25 Pa under BS 476 *Fire tests on building materials and structures*, section 31.1 *Methods for measuring smoke penetration through*

    doorsets and shutter assemblies, Method of measurement under ambient temperature conditions; or

    (b) meet the additional classification requirement of Sa when tested on BS EN 1634-3, *Fire resistance tests for door and shutter assemblies*, Part 3 – *Smoke control doors*.

3. The National classifications do not automatically equate with the equivalent classifications in the European column, therefore products cannot typically assume a European class unless they have been tested accordingly.

In addition to the guidance in this appendix, the designer may find the following references useful:

- TRADA IS 1/13 The technology of fire-resisting doorsets.
- PD 6512 Pt 1 Guide for Fire doors.
- PD 6512 Guide to the performance of Glass.
- BS 8214: 1990 Code of Practice for fire door assemblies with non-metallic leaves.
- BS 476 Pt 22 Methods for determination of the fire resistance of non-loadbearing elements of construction.
- BS 476 Pt 23 Methods for determination of the contribution of components to the fire resistance of of the structure.
- The Door and Hardware Federation (a merger of the ABHM and the D&SMA)

The choice of fire extinguisher depends on the nature of the risks likely to be encountered.

Classes of fire

Fires can generally be classified into five groups, see BS EN 2: 1992 (including amendment 2004). Fire extinguishers provided should be appropriate to the specific risks found in the premises in accordance with table D1.

Table D1 Class of fire

Class of fire	Description
Class A	Fires involving solid materials such as wood, paper or textiles
Class B	Fires involving flammable liquids such as petrol, diesel or oils
Class C	Fires involving gases
Class D	Fires involving metals
Class F	Fires involving cooking oils such as in deep-fat fryers

Types of fire extinguisher

Different types of fire extinguisher are used for different fire types. The main types are illustrated in figure D1, along with the types of fire they are not suitable for.

The recommended type and general location of fire-fighting apparatus is given in table D2.

Figure D1 Types of fire extinguisher

Main types of portable extinguishers, their uses and colour coding.

WATER	POWDER	FOAM	CARBON DIOXIDE (CO_2)
For wood, paper, textile and solid material fires	For liquid and electrical fires	For use on liquid fires	For liquid and electrical fires

DO NOT USE on liquid, electrical or metal fires	**DO NOT USE on metal fires**	**DO NOT USE on electrical or metal fires**	**DO NOT USE on metal fires**

The contents of an extinguisher is indicated by a zone of colour on the red body. Halon extinguishers are not shown since no new Halon production is permitted in the UK.

Table D2 Recommended type and location of fire-fighting apparatus

In general two 13A rated extinguishers should be provided on every floor, more if the floor area exceeds 400m². Additional extinguishers should be provided to cover different types of risk.

Type	Location
Water	Design and technology spaces Stages of every assembly hall Residential areas of boarding schools On escape routes, so that the walking distance to the nearest extinguisher does not exceed 30m
Foam or dry powder	Laboratories[1][2] Food technology[2][3] Kitchens
Wet chemical	Kitchens/Food technology for deep fat fires[3]
Foam	Boiler rooms where oil fuel is used
Carbon dioxide or dry powder[4]	Electrical switch rooms and places where live electrical equipment is known or thought to be present, eg, stage lighting control areas and ICT classrooms
Dry powder	Vehicle protection
Fire blankets	Adjacent to fire extinguisher in kitchens, laboratories, design technology practical spaces and assembly halls

Notes:

1. In some laboratories where very volatile liquids are used or fragile equipment is installed, dry powder or carbon dioxide may be preferable to foam.

2. In laboratories and food technology rooms, the capacity of extinguishers should be: for water about 9 litres capacity (13A rated), dry powder about 1.5 kg and carbon dioxide not less than 2.5 kg.

3. Where there is no fixed frying equipment, a Class F extinguisher (wet chemical) may be preferable to dry powder or foam.

4. Dry powder and carbon dioxide do not conduct electricity.

Number of extinguishers required

In simple premises, having one or two portable extinguishers of the appropriate type, readily available for use, may be all that is necessary. In more complex premises, a number of portable extinguishers may be required and they should be sited in suitable locations such as on the escape routes at each floor level. It may also be necessary to indicate the location of extinguishers by suitable signs.

Typically for the Class A fire risk, the provision of one water-based extinguisher for approximately every 200m² of floor space, with a minimum of two extinguishers per floor, will normally be adequate.

Where it is determined that there are additionally other classes of fire risk, the appropriate type, number and size of extinguisher should be provided. Further information is available in BS 5306-8.

Positioning of extinguishers

Whatever type of fire extinguishers are used, their siting in the building should, as far as is practicable, be standardised, especially if the building is on several levels. A bracket should be provided for every extinguisher and should preferably be either specially designed to prevent it being dislodged or sited in a recess. Where the wall will not support a bracket a purpose-built stand is permitted. Brackets and stands should be located so that the handle or carrying device of the extinguisher is 1m above floor level for larger extinguishers (with a total weight greater than 4kg) and 1.5m above the floor for smaller extinguishers.

Where the fire risk is not confined to a particular location, eg, Class A fires, the fire extinguishers should be positioned on escape routes, close to the exit from the room or floor, or the final exit from the building. Similarly, where the particular fire risk is specifically located, eg, flammable liquids, the appropriate fire extinguisher should be near to the hazard, so located that they can be safely used.

Extinguishers should be located in areas where they can be easily accessed by trained members of staff, but not in areas where equipment is open to misuse or vandalism.

Ideally no one should have to travel more than 30m to reach a fire extinguisher.

Further information

For further information on fire extinguishers and other portable fire-fighting equipment, refer to the RRO *Guide for Educational Premises*, Part 1 section 3.4.2 and Part 2 section 3.1.

Appendix E
Methods of measurment

Some form of measurement is an integral part of many of the provisions in this document. The following paragraphs and figures E1 to E7 show how the various forms of measurement should be made.

Travel distance

Travel distance is measured by way of the shortest route which if:

a. there is fixed seating or other fixed obstructions, is along the centre line of the seatways and gangways;

b. it includes a stair, is along the pitch line on the centre line of travel.

Width

The width of:

a. a door (or doorway) is the clear width when the door is open (see figure E1);

b. an escape route is the width at 1500mm above floor level when defined by walls or, elsewhere, the minimum width of passage available between any fixed obstructions;

c. a stair is the clear width between the walls or balustrades.

Note 1: In the case of escape routes and stairs, handrails and strings which do not intrude more than 100mm into these widths may be ignored (see figure E1).

Note 2: The rails used for guiding a stair-lift may be ignored when considering the width of a stair. However, it is important that the chair or carriage is able to be parked in a position that does not cause an obstruction to either the stair or landing.

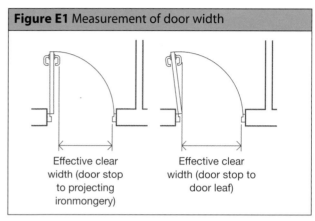

Figure E1 Measurement of door width

Effective clear width (door stop to projecting ironmongery)

Effective clear width (door stop to door leaf)

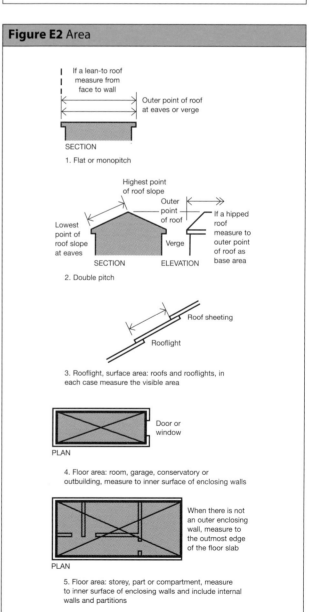

Figure E2 Area

If a lean-to roof measure from face to wall

Outer point of roof at eaves or verge

SECTION

1. Flat or monopitch

Highest point of roof slope

Outer point of roof

If a hipped roof measure to outer point of roof as base area

Lowest point of roof slope at eaves

Verge

SECTION   ELEVATION

2. Double pitch

Roof sheeting

Rooflight

3. Rooflight, surface area: roofs and rooflights, in each case measure the visible area

Door or window

PLAN

4. Floor area: room, garage, conservatory or outbuilding, measure to inner surface of enclosing walls

When there is not an outer enclosing wall, measure to the outmost edge of the floor slab

PLAN

5. Floor area: storey, part or compartment, measure to inner surface of enclosing walls and include internal walls and partitions

Figure E3 Height of building

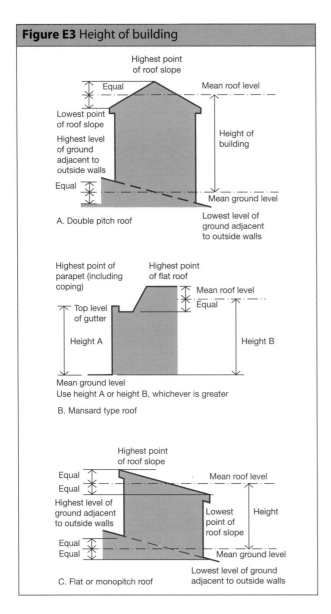

Highest point of roof slope

Equal

Lowest point of roof slope

Highest level of ground adjacent to outside walls

Equal

Mean roof level

Height of building

Mean ground level

Lowest level of ground adjacent to outside walls

A. Double pitch roof

Highest point of parapet (including coping)

Highest point of flat roof

Mean roof level

Equal

Top level of gutter

Height A

Height B

Mean ground level
Use height A or height B, whichever is greater

B. Mansard type roof

Highest point of roof slope

Equal

Equal

Highest level of ground adjacent to outside walls

Mean roof level

Lowest point of roof slope

Height

Equal

Equal

Mean ground level

Lowest level of ground adjacent to outside walls

C. Flat or monopitch roof

Figure E4 Number of storeys

To count the number of storeys in a building, or in a separated part of a building, count only at the position which gives the greatest number and exclude any basement storeys.

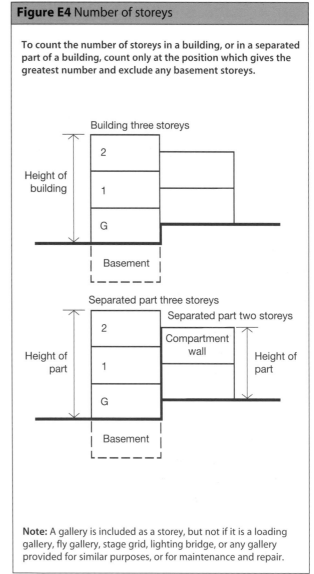

Building three storeys

Height of building

2

1

G

Basement

Separated part three storeys

Separated part two storeys

Height of part

2

1

G

Compartment wall

Height of part

Basement

Note: A gallery is included as a storey, but not if it is a loading gallery, fly gallery, stage grid, lighting bridge, or any gallery provided for similar purposes, or for maintenance and repair.

Figure E5 Height of top storey in building

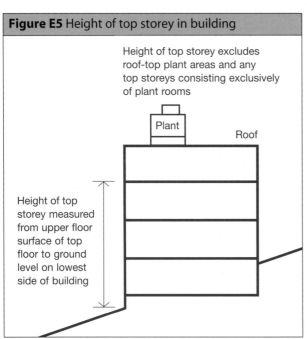

Height of top storey excludes roof-top plant areas and any top storeys consisting exclusively of plant rooms

Plant

Roof

Height of top storey measured from upper floor surface of top floor to ground level on lowest side of building

Free area of smoke ventilators

The free area of a smoke ventilator, specified here, from Approved Document B, may be measured by either:

a. the declared aerodynamic free area in accordance with BS EN 12101-2:2003 *Smoke and heat control systems. Specification for natural smoke and heat exhaust ventilators*; or,

b. The total unobstructed cross-sectional area, measured in the plane where the area is at a minimum and at right angles to the direction of air flow (see figure E6).

143

Figure E6 Free area of smoke ventilators

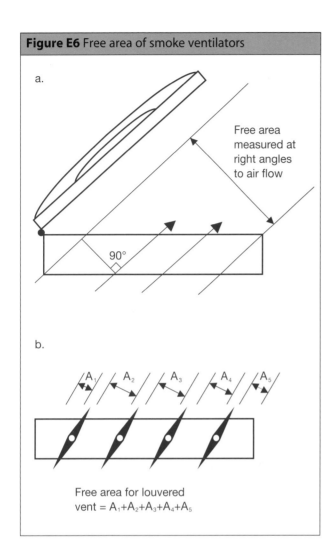

a.

Free area
measured at
right angles
to air flow

90°

b.

A_1 A_2 A_3 A_4 A_5

Free area for louvered
vent = $A_1+A_2+A_3+A_4+A_5$

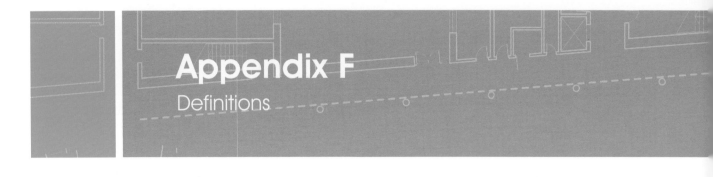
Note: Except for the items marked * (which are from the Building Regulations), these definitions which apply to BB 100 are largely taken from Approved Document B: *Fire Safety*.

Access room A room through which passes the only escape route from an inner room.

Accommodation stair A stair, additional to that or those required for escape purposes, provided for the convenience of occupants.

Alternative escape routes Escape routes sufficiently separated by either direction and space, or by fire-resisting construction, to ensure that one is still available should the other be affected by fire.

Alternative exit One of two or more exits, each of which is separate from the other.

Appliance ventilation duct A duct provided to convey combustion air to a gas appliance.

Atrium (*plural* atria) A space within a building, not necessarily vertically aligned, passing through one or more structural floors.

Note: Enclosed lift wells, enclosed escalator wells, building services' ducts and stairways are not classified as atria.

Automatic release mechanism A device which will allow a door held open by it to close automatically in the event of each or any one of the following:

a. detection of smoke by automatic apparatus suitable in nature, quality and location;

b. operation of a hand-operated switch fitted in a suitable position;

c. failure of electricity supply to the device, apparatus or switch; and

d. operation of the fire alarm system if any.

Basement storey A storey with a floor which at some point is more than 1200mm below the highest level of ground adjacent to the outside walls (however, see appendix A, table A2, for situations where the storey is considered to be a basement only because of a sloping site).

Boundary The boundary of the land belonging to the building, or, where the land abuts a road, railway, canal or river, the centreline of that road, railway, canal or river (see figure 37).

*Building Any permanent or temporary building but not any other kind of structure or erection. A reference to a building includes a reference to part of a building.

Building Control Body A term used to include both Local Authority Building Control and Approved Inspectors.

Cavity barrier A construction, other than a smoke curtain, provided to close a concealed space against penetration of smoke or flame or provided to restrict the movement of smoke or flame within such a space.

Ceiling A part of a building which encloses and is exposed overhead in a room, protected shaft or circulation space (the soffit of a rooflight is included as part of the surface of the ceiling, but not the frame. An upstand below a rooflight would be considered as a wall).

Circulation space A space (including a protected stairway) mainly used as a means of access between a room and an exit from the building or compartment.

Class 0 A product performance classification for wall and ceiling linings. The relevant test criteria are set out in appendix A.

Compartment (fire) A building or part of a building, comprising one or more rooms, spaces or storeys, constructed to prevent the spread of fire to or from another part of the same building, or an adjoining building (a roof space above the top storey of a compartment is included in that compartment. See also 'Separated part').

Compartment wall or floor A fire-resisting wall/floor used in the separation of one fire compartment from another (constructional provisions are given in 6.3).

Concealed space or cavity A space enclosed by elements of a building (including a suspended ceiling) or contained within an element, but not a room, cupboard, circulation space, protected shaft or space within a flue, chute, duct, pipe or conduit.

Corridor access A design of a building containing flats in which each flat is approached via a common horizontal internal access or circulation space which may include a common entrance hall.

Dead end Area from which escape is possible in one direction only.

Direct distance The shortest distance from any point within the floor area, measured within the external enclosures of the building, to the nearest storey exit ignoring walls, partitions and fittings, other than the enclosing walls/partitions to protected stairways.

Element of structure

a. a member forming part of the structural frame of a building or any other beam or column;

b. a loadbearing wall or loadbearing part of a wall;

c. a floor;

d. a gallery (but not a loading gallery, fly gallery, stage grid, lighting bridge, or any gallery provided for similar purposes or for maintenance and repair);

e. an external wall; and

f. a compartment wall (including a wall common to two or more buildings. However, see the guidance to B3, section 6.2.2.2, for exclusions from the provisions for elements of structure).

Emergency lighting Lighting provided for use when the supply to the normal lighting fails.

Escape lighting That part of the emergency lighting which is provided to ensure that the escape route is illuminated at all material times.

Escape route Route forming that part of the means of escape from any point in a building to a final exit.

European Technical Approval A favourable technical assessment of the fitness for use of a construction product for an intended use, issued for the purposes of the Construction Products Directive by a body authorised by a member State to issue European Technical Approvals for those purposes and notified by that member State to the European Commission.

European Technical Approvals Issuing body A body notified under Article 10 of the Construction Products Directive. The details of these institutions are published in the 'C' series of the Official Journal of the European Communities.

Evacuation lift A lift that may be used for the evacuation of people in a fire.

Exit passageway A protected passageway connecting a protected stairway to a final exit (exit passageways should be protected to the same standard as the stairway they serve).

External wall (or side of a building) Includes a part of a roof pitched at an angle of more than 70° to the horizontal, if that part of the roof adjoins a space within the building to which persons have access (but not access only for repair or maintenance).

Final exit The termination of an escape route from a building giving direct access to a street, passageway, walkway or open space and sited to ensure the rapid dispersal of persons from the vicinity of a building so that they are no longer in danger from fire and/or smoke.

Note: Windows are not acceptable as final exits

Fire damper Mechanical or intumescent device within a duct or ventilation opening which is operated automatically and is designed to prevent the passage of fire and which is capable of achieving an integrity E classification and/or an ES classification to BS EN 13501-3:2005 when tested to BS EN1366-2:1999. Intumescent fire dampers may be tested to ISO 10294-5.

Fire and smoke damper Fire damper which when tested in accordance with BS EN 1366-2:1999 meets the ES classification requirements defined in EN 13501-3:2005 and achieves the same fire resistance in relation to integrity, as the element of the building construction through which the duct passes. Intumescent fire dampers may be tested to ISO 10294-2.

Fire door A door or shutter, provided for the passage of persons, air or objects, which, together with its frame and furniture as installed in a building, is intended (when closed) to resist the passage of fire and/or gaseous products of combustion and is capable of meeting specified performance criteria to those ends. It may have one or more leaves and the term includes a cover or other form of protection to an opening in a fire-resisting wall or floor, or in a structure surrounding a protected shaft).

Fire-fighting lift A lift designed to have additional protection, with controls that enable it to be used under the direct control of the Fire and Rescue Service in fighting a fire (see sections 8.2 – 8.4).

Fire-fighting lobby A protected lobby providing access from a fire-fighting stair to the accommodation area and to any associated fire-fighting lift.

Fire-fighting shaft A protected enclosure containing a fire-fighting stair, fire-fighting lobbies and, if provided, a fire-fighting lift, together with its machine room (if relevant).

Fire-fighting stair A protected stairway communicating with the accommodation area only through a fire-fighting lobby.

Fire-resisting (fire resistance) The ability of a component or construction of a building to satisfy for a stated period of time, some or all of the appropriate criteria specified in the relevant standard test (loadbearing capacity, integrity and insulation. See appendix A).

Fire-separating element A compartment wall, compartment floor, cavity barrier and construction enclosing a protected escape route and/or a place of special fire hazard.

Fire stop A seal provided to close an imperfection of fit or design tolerance between elements or components, to restrict the passage of fire and smoke.

Gallery A floor or balcony which does not extend across the full extent of a building's footprint and is open to the floor below.

Height (of a building or storey for the purposes of BB 100) Height of a building is measured as shown in appendix E, figure E3 and height of the floor of the top storey above ground is measured as shown in appendix E, figure E5.

Inner room Room from which escape is possible only by passing through another room (the access room).

Material of limited combustibility A material performance specification that includes non-combustible materials and for which the relevant test criteria are set out in appendix A.

Means of escape Structural means whereby (in the event of fire) a safe route or routes is or are provided for persons to travel from any point in a building to a place of safety.

Measurement Width of a doorway, area, cubic capacity, height of a building and number of storeys, see appendix E, figures E1 to E6. For travel distance and, width of escape route and a stair, see appendix E.

Non-combustible material The highest level of reaction to fire performance. The relevant test criteria are set out in appendix A.

Notional boundary A boundary presumed to exist between buildings on the same site (see section 7.3, figure 38).

Open spatial planning The internal arrangement of a building in which more than one storey or level is contained in one undivided volume, eg, split-level floors. For the purposes of this document there is a distinction between open spatial planning and an atrium space.

Perimeter (of a building) The maximum aggregate plan perimeter, found by vertical projection onto a horizontal plane (see section 8.3, figure 43).

Pipe (for the purposes of section 6.5) Includes pipe fittings and accessories and excludes a flue pipe and a pipe used for ventilating purposes (other than a ventilating pipe for an above around drainage system).

Places of special fire hazard Oil-filled transformer and switch-gear rooms, boiler rooms, storage space for fuel or other highly flammable substances and rooms housing a fixed internal combustion engine. Additionally in schools the list includes laboratories, technology rooms with open heat sources, kitchens and stores for PE mats or chemicals.

Platform floor (access or raised floor) A floor supported by a structural floor, but with an intervening concealed space which is intended to house services.

Protected circuit An electrical circuit protected against fire.

Protected corridor/lobby A corridor or lobby which is adequately protected from fire in adjoining accommodation by fire-resisting construction.

Protected shaft A shaft which enables persons, air or objects to pass from one compartment to another and which is enclosed with fire-resisting construction.

Protected stairway A stair discharging through a final exit to a place of safety (including any exit passageway between the foot of the stair and the final exit) that is adequately enclosed with fire-resisting construction.

Relevant boundary The boundary which the side of the building faces, (and/or coincides with) and which is parallel, or at an angle of not more than 80°, to the side of the building (see section 7.3, figure 37). A notional boundary can be a relevant boundary.

Rooflight A dome light, lantern light, skylight, ridge light, glazed barrel vault or other element intended to admit daylight through a roof.

Room (for the purposes of B2) An enclosed space within a building that is not used solely as a circulation space (the term includes not only conventional rooms, but also cupboards that are not fittings and large spaces such as warehouses and auditoria. The term does not include voids such as ducts, ceiling voids and roof spaces).

School A place of education for children older than 2 and younger than 19 years. Includes nursery schools, primary schools and secondary schools as defined in the Education Act 1996.

Self-closing device A device which is capable of closing the door from any angle and against any latch fitted to the door.

Note: Rising butt hinges which do not meet the above criteria are acceptable where the door is in a cavity barrier.

Separated part (of a building) A form of compartmentation in which a part of a building is separated from another part of the same building by a compartment wall. The wall runs the full height of the part and is in one vertical plane (see section 6.3.3.3 and appendix E, figure E4).

Single storey building A building consisting of a ground storey only (a separated part which consists of a ground storey only, with a roof to which access is only provided for repair or maintenance, may be treated as a single storey building). Basements are not included in counting the number of storeys in a building (see appendix E).

Site (of a building) is the land occupied by the building, up to the boundaries with land in other ownership.

Smoke alarm A device containing within one housing all the components, except possibly the energy source, necessary for detecting smoke and giving an audible alarm.

Storey Includes:

a. any gallery; and

b. a roof, unless it is accessible only for maintenance and repair.

Storey exit A final exit, or a doorway giving direct access into a protected stairway, fire-fighting lobby, or external escape route.

Suspended ceiling (fire-protecting) A ceiling suspended below a floor, which contributes to the fire resistance of the floor. Appendix A, table A3, classifies different types of suspended ceiling.

Technical specification A standard or a European Technical Approval Guide. It is the document against which compliance can be shown in the case of a standard and against which an assessment is made to deliver the European Technical Approval.

Thermoplastic material See appendix A

Travel distance (unless otherwise specified) The actual distance to be travelled by a person from any point within the floor area to the nearest storey exit, having regard to the layout of walls, partitions and fittings.

Unprotected area In relation to a side or external wall of a building means:

a. window, door or other opening; and

   Note: Windows that are not openable and are designed and glazed to provide the necessary level of fire resistance and recessed car parking areas shown in figure F1, need not be regarded as an unprotected area.

b. any part of the external wall which has less than the relevant fire resistance set out in section 7.2; and

c. any part of the external wall which has combustible material more than 1mm thick attached or applied to its external face, whether for cladding or any other purpose (combustible material in this context is any material which does not have a Class 0 rating.)

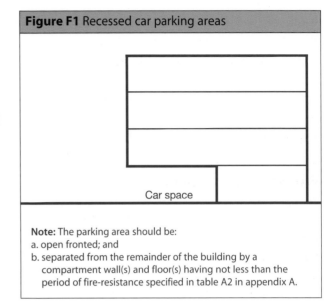

Figure F1 Recessed car parking areas

Car space

Note: The parking area should be:
a. open fronted; and
b. separated from the remainder of the building by a compartment wall(s) and floor(s) having not less than the period of fire-resistance specified in table A2 in appendix A.

Appendix G
Fire safety information

Regulation 16B requires that, where building work involves the erection or extension of a relevant building, or a relevant change of use of a building, fire safety information shall be given to the responsible person at the completion of the project or when the building or extension is first occupied.

- 'Fire safety information' means information relating to the design and construction of the building or extension, and the services, fittings and equipment provided in or in connection with the building or extension which will assist the responsible person to operate and maintain the building or extension with reasonable safety.

- A 'relevant building' is a building to which the Regulatory Reform (Fire Safety) Order 2005 applies (S.I. 2005/1541; see article 6) , or will apply after the completion of building work.

- A 'relevant change of use' is a material change of use where, after the change of use takes place, the Regulatory Reform (Fire Safety) Order 2005 will apply, or continue to apply, to the building.

- 'Responsible person' has the meaning given in article 3 of the Regulatory Reform (Fire Safety) Order 2005.

This appendix is only intended as a guide as to the kind of information that should be provided. For clarity the guidance is given in terms of simple and complex buildings, however the level of detail required will vary from building to building and should be considered on a case by case basis.

Simple buildings

For most buildings basic information on the location of fire protection measures may be all that is necessary. An as-built plan of the building should be provided showing:

a. escape routes;

b. compartmentation and separation (ie, location of fire separating elements, including cavity barriers in walk-in spaces);

c. fire doors, self-closing fire doors and other doors equipped with relevant hardware (eg, panic locks);

d. locations of fire and/or smoke detector heads, alarm call-points, detection/alarm control boxes, alarm sounders, fire safety signage, emergency lighting, fire extinguishers, dry or wet risers and other fire-fighting equipment and location of hydrants outside the building;

e. any sprinkler system(s), including isolating valves and control equipment;

f. any smoke-control system(s) (or ventilation system with a smoke-control function), including mode of operation and control systems;

g. any high-risk areas (eg, heating machinery);

h. specifications of any fire safety equipment provided, in particular any routine maintenance schedules;

i. any assumptions in the design of the fire safety arrangements regarding the management of the building; and

j. any provision incorporated into the building to facilitate the evacuation of disabled people. This information can then be used when designing suitable Personal Emergency Escape Plans.

Figure G1 Example of a line drawing showing general fire safety precautions

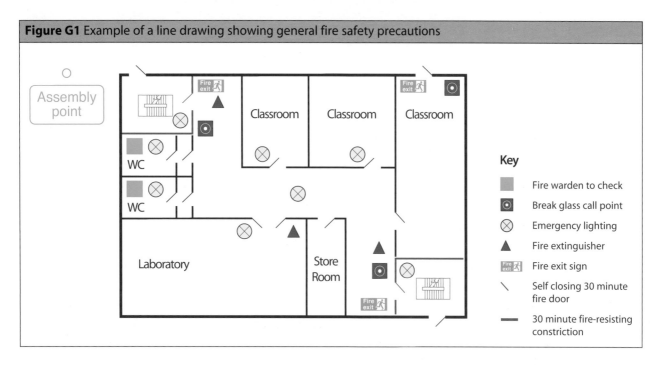

Complex buildings

For more complex buildings a more detailed record of the fire safety strategy and procedures for operating and maintaining any fire protection measures of the building will be necessary. Further guidance is available in BS5588-12:2004 *Fire precautions in the design, construction and use of buildings: Managing fire safety* (Annex A Fire Safety Manual.) These records should include:

a. the fire safety strategy, including all assumptions in the design of the fire safety systems (such as fire load). Any risk assessments or risk analysis;

b. all assumptions in the design of the fire safety arrangements regarding the management of the building;

c. escape routes, escape strategy (eg, simultaneous or phased) and muster points;

d. details of all passive fire safety measures, including compartmentation (ie, location of fire separating elements), cavity barriers, fire doors, self-closing fire doors and other doors equipped with relevant hardware (eg, electronic security locks), duct dampers and fire shutters;

e. fire detector heads, smoke detector heads, alarm call-points, detection/alarm control boxes, alarm sounders, emergency communications systems, CCTV, fire safety signage, emergency lighting, fire

extinguishers, dry or wet risers and other fire-fighting equipment, other interior facilities for the Fire and Rescue Service, emergency control rooms, location of hydrants outside the building, and other exterior facilities for the Fire and Rescue Service;

f. details of all active fire safety measures, including:

- sprinkler system(s) design, including isolating valves and control equipment; and

- smoke-control system(s) (or HVAC system with a smoke-control function) design, including mode of operation and control systems;

g. any high-risk areas (eg, heat bay equipment) and particular hazards;

h. as-built plans of the building showing the locations of the above;

i. specifications of any fire safety equipment provided, including operational details, operators manuals, software, system zoning and routine inspection, testing and maintenance schedules. Records of any acceptance or commissioning tests;

j. any provision incorporated into the building to facilitate the evacuation of disabled people; and

k. any other details appropriate for the specific building.

Appendix H
Fire risk assessment

Risk assessment and cost-benefit analysis tools

The tools are available on the enclosed CD-Rom and updates will be provided on the DCSF website www.teachernet.gov.uk/fire These are, a simple qualitative risk assessment tool, and a more comprehensive, quantitative, risk assessment and cost-benefit analysis tool.

Simple risk tool

This risk assessment tool is intended for designers, architects, fire safety engineers or others who wish to assess the need for sprinklers in their proposed school building. The tool comes in two slightly different forms, one for existing sites (refurbishment or extension) and one for new schools. As recommended by BB 100, this tool should be used as a guide to assist in the risk-based decision-making process. The weightings assigned and scoring in the current version are those agreed by the DCSF Advisory Group at the time of release.

This survey and risk assessment will allow the user to determine:

- type and scale of risk;
- trends or patterns in fire incidents involving the school;
- fire safety or fire protection measures required; and
- efficiency of your chosen fire safety measures.

Completing the survey should take 15 – 30 minutes. The user answers questions in the four-part assessment that covers:

- incidence of arson and fire;
- environment and buildings;
- effectiveness of fire safety and fire protection measures; and
- consequences of a fire.

Each question is considered in relation to the individual building and scored accordingly. For example, in one particular question, if it is believed that out-of-hours use is reducing the risks to the school building, then score low.

Figure H1 Example question from the simple risk assessment tool

In the example below, if your school has had no cases of arson reported in the last 5 years then the risk would be perceived as low and a score of 0 would be recorded:

Low Risk	0	1	2	3	4	5	High Risk

1. Arson (in the last 5 years)

No cases of arson within school grounds	⦿0	○1	○2	○3	○4	○5	Arson common within school grounds

If, however, your school had a few cases of arson over then last five years, then the score could be three of higher:

Low Risk	0	1	2	3	4	5	High Risk

1. Arson (in the last 5 years)

No cases of arson within school grounds	○0	○1	○2	⦿3	○4	○5	Arson common within school grounds

Figure H2 Score calculation and interpretation within the simple risk assessment tool

Score

Part 1	Incidence of arson (fire)	0	
Part 2	Environment and buildings	0	0
Part 3	Fire safety or fire protection measures	0	
Part 4	Cosequences of a fire	0	0
	Total	0	

Scoring

Proposed overall scoring		Proposed scoring Parts 1 and 2		Proposed scoring Parts 3 and 4	
Low risk	0 – 30	Low risk	0 – 15	Low risk	0 – 20
Average risk	31 – 80	Average risk	16 – 35	Average risk	21 – 40
High risk	81 – 200	High risk	36 – 65	High risk	41 – 135

Overall Score

Low risk
The fire safety and fire protection survey and risk assessment indicates your school is at low level of risk. Sprinklers may be beneficial.

Medium risk
The fire safety and fire protection survey and risk assessment indicates your school is at an average level of risk. A sprinkler system is desirable.

High risk
The fire safety and fire protection survey and risk assessment indicates your school is at a high level of risk. Sprinklers should be provided.

Based on the level of threat, the assessment recommends appropriate fire safety or fire protection measures for the school.

The tool addresses all types of school building - including those used only by staff. Some responses may be appropriate only for very specific types of building.

Note that this tool is concerned with life safety, building and property protection, and, to a lesser extent, environmental protection – some questions may implicitly address only one of these.

Part 1 Incidence of fire (likelihood, based on historical cases)

This section assesses the type, scale, and patterns of incidents that have occurred, primarily in the last 5 years.

This part of the risk assessment can be based on a log of reported incidents. If the school does not yet have a reporting procedure, a more subjective assessment will have to be made, possibly based on discussion with the local community fire safety or fire prevention officer.

Part 2 Environment and buildings (exposure)

This section of the survey assesses the environmental and building factors that contribute to the exposure of the school to the risk of fire, ie, its vulnerability.

Part 3 Fire safety and fire protection measures

This section assesses the effectiveness of the fire safety or fire prevention measures employed.

Part 4 Consequences/ impact of fire

This section assesses the effect of a fire on the pupils, school users and the school building.

The items in parts 2 and 3 can (for the most part) be assessed by a walk-through survey of the school.

Each item in the assessment is graded from 0 to 5 points. A score of 0 indicates a low fire risk, 5 a high fire risk.

Comment pop-ups are provided to guide users to reply appropriately.

New Build refers to school projects where there is no history of a school building on or very near to the site.

Existing site refers to school projects on or near to the site of a previous school building.

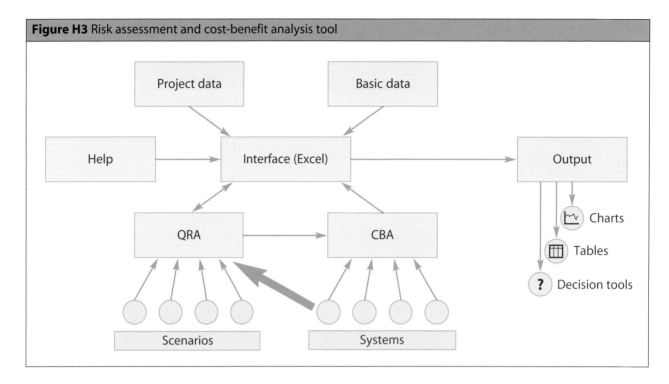

Figure H3 Risk assessment and cost-benefit analysis tool

Risk assessment and cost-benefit analysis tool

BRE have developed, on behalf of DCSF, a spreadsheet-based CBA tool. It consists of a number of interlinked modules (represented by separate sheets), which will each be described briefly.

The risk assessment (QRA) module considers the frequencies and consequences of different fire scenarios. Fires can start in different room types, eg, classrooms, cloakrooms, corridors, main hall, labs, offices and stores, and have probabilities of growing to different sizes:

- minor = item first ignited;
- severe = fire damage to room of origin and smoke damage beyond;
- major = fire damage beyond room of origin; and
- catastrophic = fire damage to most of the building, total write-off.

The consequences of the different fire scenarios are quantified in terms of

- deaths and injuries;
- damage to building and contents;
- closure of facilities; and
- environmental impact (this is a bit subjective, although usually a small component of the overall risk).

All of these are converted to monetary values.

The CBA module calculates the costs, per year, of the various fire safety systems (the tool is not just restricted to sprinklers). Up-front installation costs are converted to annual equivalents using Treasury discounting rules. Various benefits, not related to risk reduction (eg, insurance premiums) are also calculated by this module.

The CBA and QRA modules interact: the package of systems can either reduce the probabilities (frequencies) of scenarios, or their consequences, or both. For example, sprinklers would not prevent fires from starting, but they would significantly reduce the probability that a small fire can grow to a large one.

The basic data module just supplies the monetary conversion factors for the risk components (eg, a life ≈ £1.5m) and a few other constants.

The project data defines the school by the number of rooms of different types, and the total area of these.

Output takes the form of various charts, tables, etc.

Help is provided by the user manual, and also pop-up comments in various cells that explain the numbers that are input or calculated there.

British Standards

A complete list of British Standards can be obtained from the British Standards Institute website. Alternatively, relevant standards may also be found in Approved Document B (volume 2, Appendix H).

DCSF Building Bulletins and other publications

Building Bulletin 7: *Fire and the design of educational buildings (sixth edition)*, HMSO, 1992 (withdrawn).

Building Bulletin 70: *Maintenance of Mechanical Services. Maintenance and Renewal in Educational Buildings*, ISBN-0-11-270717-3.

Building Bulletin 93: *Acoustic design of schools*, ISBN 0 11 271105 7.

Building Bulletin 95: *Schools for the Future. Designs for Learning Communities.*

Building Bulletin 98: *Briefing Framework for Secondary School Projects.*

DfEE Managing School Facilities Guides; Guide 4, *Improving Security in schools*, 1996.

DfEE Managing School Facilities Guides; Guide 6 *Fire Safety*, 2000.

Schools Building and Design Unit; *Key Design Guidance for Schools: Access to information for school design*, April 2003 (updated August 2004).

Standard specifications, layouts and dimensions (SSLDs):
Sprinklers in Schools
Internal Doorsets in Schools
Stairways in Schools
Floor finishes in schools
Roof coverings in schools

Health And Safety: *Responsibilities And Powers. Organisation & Management*, DfES/0803/2001.

CLG Approved documents and other publications

Approved documents are available a free downloads from CLG:

http://www.planningportal.gov.uk/england/professionals/en/1115314110382.html

Hard copies can be obtained as a priced publication available from RIBA Bookshops Mail Order, 15 Bonhill Street, London, EC2P 2EA.

Approved Document B: *(Fire safety) – Volume 2 – Buildings other than dwellinghouses (2006 Edition)*, published 18 December 2006, ISBN: 978 1 85946 262 1.

Approved Document E: *Resistance to the passage of sound*, published 27 April 2006, ISBN: 978 1 85946 204 1.

Approved Document K: *Protection from falling, collision and impact*, published 3 May 2006, ISBN: 978 1 85946 210 2.

Approved Document M: *Access to and Use of Buildings*, published 3 May 2006, ISBN: 978 1 85946 211 9.

Approved Document N: *Glazing – safety in relation to impact, opening and cleaning*, published 3 May 2006, ISBN: 978 1 85946 212 6.

Fire Safety Risk Assessment – Educational Premises, published 5 June 2006, ISBN: 978 1 85112 819 8.
http://www.communities.gov.uk/index.asp?id=1162108

Arson Control Forum. Research Bulletin no. 10. *Survey of school fires*. December 2006.

Fire Statistics, United Kingdom, CLG, Fire Statistics and Research Division. Various years.

Other government guidance

Safety signs and signals: the Health and Safety (Safety Signs and Signals) Regulations. Guidance on Regulations, 1996, HSE publication, ISBN 0717608700.

BRE and Loss Prevention Council reports

BRE 454: *Multi-storey timber frame buildings – a design guide*, 2003, ISBN: 8608 605 3.

BRE Digest 208: *Increasing the fire resistance of existing timber floors*, 1988, ISBN: 978 8608 359 7.

BRE Report: *Design methodologies for smoke and heat exhaust ventilation*, BR 368, 1999, ISBN: 978 8608 289 7.

BRE Report: *External fire spread: Building separation and boundary distances*, BR 187, BRE 1991.

BRE Report: *Fire performance of external thermal insulation for walls of multi storey buildings*, BR 135.

BRE Report: *Fire safety of PTFE-based materials used in buildings*, BR 274, BRE 1994.

BRE Report: *Guidelines for the construction of fire-resisting structural elements*, BR 128, BRE 1988.

Loss Prevention Standard LPS 1162: Issue 3.1: 16/09/05 *Requirements and Tests for LPCB Certification of Fire Dampers 62*.

Loss Prevention Standard LPS 1175: Issue 5.2: *Requirements and Testing Procedures for the LPCB Approval and Listing of Burglary Resistant Building Components, Strongpoints and Security Enclosures S1175 – Doors, windows, shutters other façade elements and enclosures*.

Loss Prevention Standard LPS 1214: Issue 2.1: September 2005. *Specification for testing and classifying physical protection devices for personal computers and similar equipment*.

Loss Prevention Standard LPS 1602: Issue 1: *Requirements for LPCB Approval and Listing of Intruder Alarm Movement Detectors*.

Loss Prevention Standard LPS 1603: Issue 1: *Requirements for LPCB Approval and Listing of Intruder Alarm Control and Indicating Equipment*.

LPC Design Guide for the Fire Protection of Buildings, 2000 (FPA Design Guide for the Fire Protection of Buildings).

Law, M., *Heat Radiation from Fires and Building Separation*, Fire Research. Technical Paper No. 5, 1963, Her Majesty's Stationery Office, London.

Industry guidance

A guide to best practice in the specification and use of fire-resistant glazed systems, published by the Glass and Glazing Federation, 44– 48 Borough High Street, London, SE1 1XB www.ggf.org.uk

Association for Specialist Fire Protection publications are freely available from www.asfp.org.uk

ASFP *Blue Book: fire-resisting ductwork,* ISBN: 1 87040 926 4.

ASFP *Grey Book: Fire and smoke resisting dampers,* ISBN: 1 87040 924 8.

ASFP *Red Book: Fire Stopping and Penetration Seals for the Construction Industry,* ISBN: 1 87040 923 X.

ASFP *Yellow Book: Fire protection for structural steel in buildings,* 4th Edition, ISBN: 1 87040 925 6.

ASFP: *Ensuring best practice for passive fire protection in buildings,* ISBN: 1 87040 919 1

CIBSE Guide E: *Fire engineering*, 2003. ISBN 10: 1-903287-31-6. ISBN 13: 978-1-903287-31-6.

Code of practice for fire-resisting metal doorsets, published by the DSMA (Door and Shutter Manufacturers' Association), 1999.

Code of Practice, *Hardware for Timber Fire and Escape Doors,* Issue 1, Hardware Industry Federation, November 2000.

Gas Installations for Educational Establishments, IGE/UP/11, 2004. Published by the Institution of Gas Engineers (www.igem.org.uk).

Guidance Notes on Gas Safety in Educational Establishments, IM/25, 1989, available from DfES Schools Building and Design Unit or the Institute of Gas Engineers. Now superseded by the Institute of Gas Engineers UP11 *Gas installations for Educational Establishments, 2004.*

How to combat arson in schools, published by the Arson Prevention Bureau, 2003.

Intumescent Fire Seals Association, IFSA Information Sheet No 2 *The role of intumescent materials in timber and metal based fire-resisting glazing systems, 1998.*

Intumescent Fire Seals Association, the IFSA Code 1999 *Sealing apertures and service penetrations to maintain fire resistance.*

SCI P 197 *Designing for structural fire safety: A handbook for architects and engineers, 1999,* ISBN: 1 85942 074 5.

SCI Publication 288 *Fire safe design: A new approach to multi-storey steel-framed buildings,* Second Edition 2000, ISBN: 1 85942 169 5.

Secured by Design (SBD), ACPO, see http://www.securedbydesign.com/apply/membership.aspx

Sprinklers for Safety: Use and Benefits of Incorporating Sprinklers in Buildings and Structures, BAFSA 2006, ISBN: 0 95526 280 1.

The design and protection of new school buildings and sites, Zurich Municipal Insurance, 2005. Download from www.zurichmunicipal.com

Timber Fire-Resisting Doorsets: maintaining performance under the new European test standard, 2002, published by TRADA, ISBN 1900510359.